Springer
Berlin
Heidelberg
New York
Barcelona
Budapest
Hong Kong
London
Mailand
Paris
Tokyo

European Science Foundation

The **European Science Foundation** is an association of its 55 member research councils, academies, and institutions devoted to basic scientific research in 20 countries. The ESF assists its Member Organisations in two main ways: by bringing scientists together in its Scientific Programmes, Networks and European Research Conferences, to work on topics of common concern; and through the joint study of issues of strategic importance in European science policy.

The scientific work sponsored by ESF includes basic research in the natural and technical sciences, the medical and biosciences, the humanities and social sciences.

The ESF maintains close relations with other scientific institutions within and outside Europe. By its activities, ESF adds value by cooperation and coordination across national frontiers and endeavours, offers expert scientific advice on strategic issues, and provides the European forum for fundamental science.

This volume arises from the work of the ESF Scientific Programme on Regional and Urban Restructuring in Europe (RURE).

Further information on ESF activities can be obtained from:

European Science Foundation
1, quai Lezay-Marnésia
F-67080 Strasbourg Cedex
France

Tel. (+33) 88 76 71 00
Fax (+33) 88 37 05 32

Ray Hudson   Eike W. Schamp (Eds.)

Towards a New Map of Automobile Manufacturing in Europe?

New Production Concepts and Spatial Restructuring

With Contributions by
Ash Amin   Sergio Conti   Peter Dicken   Aldo Enriette
Joao Ferrao   Ray Hudson   Anders Malmberg   David Sadler
Julien Savary   Eike W. Schamp   Mario Vale

Springer

Professor Ray Hudson

University of Durham
Department of Geography
DHI 3LE Durham, U.K.

Professor Eike W. Schamp

Johann-Wolfgang-Goethe-Universität
Institut für Wirtschafts- und Sozialgeographie
Dantestraße 9
D-60325 Frankfurt am Main, Germany

ISBN-13: 978-3-642-79473-5 e-ISBN-13: 978-3-642-79471-1
DOI: 10.1007/978-3-642-79471-1

Softcover reprint of the hardcover 1st edition 1995

SPIN: 10425838 30/3136 - 5 4 3 2 1 0 - Printed on acid-free paper

Contents

Tables

Figures

List of Contributors

Professor Ash Amin
CURDS, University of Newcastle upon Tyne,
Newcastle upon Tyne NE1 7RU, U.K.
Professor Sergio Conti
Dipartimento Interateneo Territorio, Politecnico e Università di Torino,
Sede di C.so, Palestro 5, I-10122 Torino, Italy
Professor Peter Dicken
School of Geography, University of Manchester,
Oxford Road, Manchester M13 9PL, U.K.
Dr. Aldo Enrietti
Dipartimento di Economia, Università di Torino,
Via Po 53, I-10100 Torino, Italy
Dr. João Ferrão
Instituto de Ciencias Sociais, Universidade de Lisboa, Av. Forças Armadas,
Edif. ISCTE, 1600 Lisboa, Portugal
Professor Ray Hudson
Department of Geography, University of Durham, South Road,
Durham DH1 3LE, U.K.
Dr. Anders Malmberg
Department of Social and Economic Geography, Uppsala University,
Box 554, S-751 22 Uppsala, Sweden
Dr. David Sadler
Department of Geography, University of Durham, South Road,
Durham DH1 3LE, U.K.
Dr. Julien Savary
Laboratoire d'Etudes et de Recherches sur l'Economie de la Production
(LEREP), Université des Sciences Sociales de Toulouse, Place Anatol France,
F-31042 Toulouse Cedex, France
Professor Eike W. Schamp
Institut für Wirtschafts- und Sozialgeographie,
Johann Wolfgang Goethe Universität Frankfurt/Main,
Dantestraße 9, D-60054 Frankfurt/Main, Germany
Dr. Mario Vale
Instituto Nacional de Investigação Cientifica, Centro de Estudios Geográficos,
Faculdade de Letras, Cidade Universitária, 1699 Lisboa Codex, Portugal

Chapter 1
New Challenges to the Automobile Production Systems in Europe

Peter Dicken [1], Ray Hudson [2] and Eike W. Schamp [3]
[1] Department of Geography, University of Manchester, United Kingdom
[2] Department of Geography, University of Durham, United Kingdom
[3] Institute of Economic and Social Geography, Johann Wolfgang Goethe University, Frankfurt am Main, Germany

1.1 Introduction

The European automobile industry has never before manufactured as many cars as it does today[1]. Equally, private households in western Europe have never possessed as many cars as they currently do, whether measured absolutely or relatively (in automobiles per thousand inhabitants). In 1991, in terms of the number of registrations, the European Community market was almost comparable in size to that of the USA, with 140 and 145 million registrations, respectively. But the market in the larger European Union will greatly exceed that in the USA and dwarf the smaller Japanese market. With 37 million registrations in 1991, the Japanese market was approximately the same size as that in West Germany alone. Future prospects look good in view of the rapidly emerging east European markets. It would clearly seem that the earlier economic shocks of the 1970s have been splendidly overcome during the 1980s in Europe.

And yet many observers claim that the European automobile sector is in crisis in the early 1990s. This is a matter for serious concern as the automobile industry and its related sectors are crucial to national economic performance in

[1]The chapters in this book deal only with cars and related new niche market products such as multi-purpose vehicles and off-the-road leisure vehicles but not buses or trucks.

the larger European countries, such as France, Germany, Italy, Spain and the UK. It is in these countries that the automobile was invented around one hundred years ago. The auto industry has since grown to become a powerful key economic sector in the last five decades (Laux, 1992). But it is a sector that has also become a prime example of an old industry maturing in industrialised countries, characterised by lower productivity compared to newer global competitors (Womack et al., 1990), decreasing employment and, sometimes, violent fluctuations in sales. In addition to such alarming signs, which are significant in themselves, most observers in the social sciences now agree that, underlying them, there is a more fundamental, radical change in the rationale of modern capitalist production which has been taking place since the 1970s. Such changes were manifest in the automobile sector at an early stage - the very sector which gave its name to the previously dominant 'Fordist' production rationale. For several decades, this 'Fordist' regime had provided an established and stable framework for the organisation and operation of production. The outcome of the recent and still ongoing turbulent changes is, however, not yet fully understood. Will it be the final sunset of a maturing and no longer sunrise sector? Does it signal the decline of old industrialised Europe in general? And what does it mean for regions and localities in Europe?

But as if these structural changes were not enough, political and economic conditions have also radically altered in Europe in the early 1990s. The continent is in the midst of a long-term process of transformation of its economic structure, the end of which is not yet in sight. First, the project to complete the Single European Market tried to integrate the then existing national markets of EC members into a real and genuinely unified market. This was expected to lead both to a rejuvenation of production structures and to an increased level of private consumption. But in the course of this rebuilding of the European economy, which had already resulted in an increasing concentration of larger firms for a larger market, came the dramatic breakdown of the state socialist system in Eastern Europe. This could as easily be considered as a kind of destabilizing political hazard as it could be seen as an economic opportunity. Nevertheless, it resulted in an immediate opening of new markets for automobiles and other consumer goods which, in the short term, favoured the West European production system. In the long term, however, the transition of East European countries into mixed or market economies and their integration into trade with the European Union will significantly alter conditions for the spatial allocation of production over the greater European space.

In this introductory chapter we seek to establish a context for the rest of the book in two ways. First, we seek to locate automobile production and consumption within Europe in its broader international context. Europe is part of an increasingly globalized economy, especially in sectors such as automobiles, and consequently production within Europe reflects its relationships with that wider economy. The changing geography of production within Europe is in part explicable as a consequence of these relationships while these wider relationships are also shaped in important ways by changes in the automobile sector in Europe. In short, changes at the regional scale of Europe, at the local scale within Europe and at the global scale are mutually determining. Second, recognizing that there is a variety of ways of conceptualizing this changing organization of production and its geographies, we sketch out a production systems approach. This encompasses both the strategies of automobile manufacturers and their relationships with other companies within the social division of labour of the production system defined in relation to this sector. This forms a starting point for subsequent more detailed discussion as we point to issues that are considered in greater depth in following chapters, and situate the rest of the book in relation to this conceptual framework.

1.2 The European Automobile Industry in Global Context

The chapters of this book focus primarily on the dynamics of the automobile industry within Europe and on the European strategies of the leading European-based automobile producers. It is important, however, to set these strategies within a broader context. The automobile industry is one of the most global of contemporary industries; it is an industry dominated by giant companies which are increasingly, although to varying degrees, organising their production along transnational lines (Dicken, 1992b). Competition within the automobile industry is truly global even though the competitive struggles themselves are fought out on specific regional and national battlefields. Between the mid-1970s and the mid-1980s, the major battleground for these so-called 'Car Wars' was North America. More recently, the battle for competitive advantage between the major producers has shifted to Europe - but to a Europe which now looks very different from the Europe of the mid-1980s. At that time, the moves towards the Single European Market had only just begun (although, as argued by Dicken, 1992a), some of the major automobile firms had already put in place pan-European operations) and there were no signs of the political upheavals which were to break out in Eastern Europe and the former Soviet Union. At that time, the European automobile industry was, to all intents and purposes, a western European industry. The position in the 1990s is very different. Not only is the European automobile industry itself part of a much enlarged economic and political space but also political developments in North America towards the creation of a North American Free Trade Agreement (NAFTA) will undoubtedly have repercussions for the global automobile industry as a whole as well as for the European industry. At the same time, new players have begun to emerge from some of the newly industrialized economies of East and South East Asia, notably from South Korea and, to a lesser extent, Malaysia.

1.2.1 The global significance of the European automobile system

It is not difficult to understand why the competitive focus in the automobile industry has shifted to Europe. More automobiles are produced and sold in Europe than in any other region of the world (Table 1.1). In 1992, both automobile production and automobile sales in western Europe amounted to around 38 per cent of the world total. Although these levels were down from a peak in 1989/1990, this was in the context of a world-wide fall after the boom years of the late-1980s. The close correspondence between production and sales suggests that, in aggregate terms, the western European industry is more or less in balance. In fact, there are enormous differences in the ratio of production and sales - and, therefore, of imports and exports - between individual European countries (Dicken, 1992a).

Almost 80 per cent of western Europe's automobile trade is intra-regional and constitutes an intricate network of product (and component) flows which reflect both the sourcing and marketing strategies of the major automobile producers, national and transnational. Individual country automobile trade deficits and surpluses are the product of a complex set of circumstances: varying combinations of arm's length trade and of the tied imports arising from intra-European sourcing strategies of the transnational companies with European production bases.

As Figure 1.1. shows, the major inflow of automobiles to Europe is from Japan but this is more than balanced, in aggregate terms, by an outflow of

Table 1.1 Automobile production and sales in Western Europe, 1991

	Percent of world total	
	Production	Sales
Western Europe	38.2	37.5
European Community	38.1	34.6
West Germany	14.7	10.5
France	9.4	6.1
Spain	5.3	2.6
Italy	4.7	6.4
United Kingdom	3.5	4.9
North America	18.5	27.9
United States	15.9	n.a.
Canada	2.7	n.a.
Asia		
Japan	29.4	14.2
South Korea	3.5	2.0

Source : SMMT 1992; Financial Times 11 September 1991

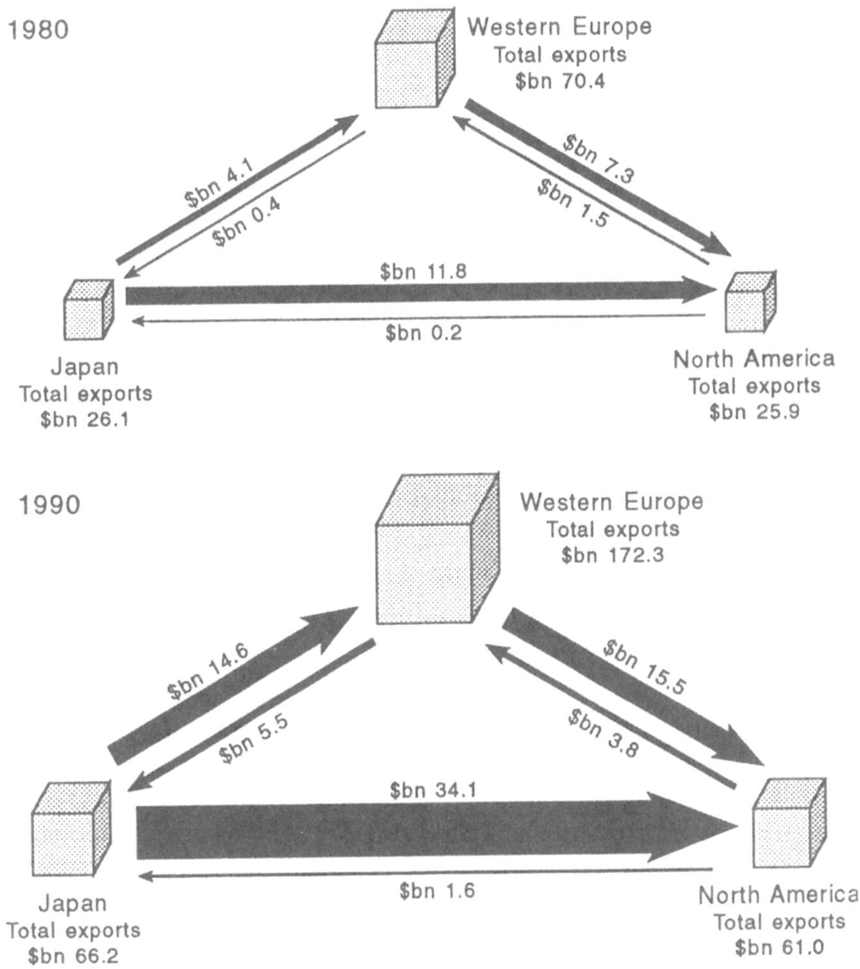

Fig. 1.1 Global trade in automobiles, 1980 and 1990.

mostly high-value, luxury automobiles from Europe to North America. In contrast, Japan has a massive automobile trade surplus with both western Europe and North America while North America itself has an even larger automobile trade deficit with both Japan and western Europe. It is especially interesting to compare western Europe's international trade in automobiles (excluding intra-regional trade) with that of Japan between 1980 and 1990. Table 1.2 reveals that, taking Japan's trade as 100 in each year, western Europe's position deteriorated from an index value of 86.2 in 1980 to a value of 55.4 in 1990.

A further measure of the global position of western Europe is reflected in its importance as a market to the automobile companies of Japan and the United States. In 1991, western European consumers absorbed almost one-third (31.9

Table 1.2 External trade in automobiles : Japan, Western Europe and North America

	1990	1980
Japan	100.0	100.0
Western Europe	55.4	86.2
North America	20.7	31.4

Source calculated from GATT (1992), volume 2, Table IV, 42

per cent) of Japan's total automobile exports whereas North America absorbed 45 per cent of Japanese car exports. Clearly, western Europe is still substantially less important as an export market (and as a production base) for Japanese automobile producers although it has to be recognised that the inflow of Japanese cars into both western Europe and North America has been heavily controlled through a series of so-called 'voluntary' export restraint agreements. No such restrictions have affected automobile exports to Europe from North America but, even so, western Europe accounts for only 11 per cent of the total.

1.2.2 Europe's position in the global strategies of the major automobile producers

Europe's position within the global network of automobile trade and on the production map are two closely related facets of Europe's position within the global automobile industry. The other major dimension to be addressed here - and one that is central to the entire book - is the region's relative significance in the overall competitive strategies of the major automobile producers themselves. Individual chapters examine these strategies in considerable detail. Hence, in setting the scene here, the aim is merely to construct a broad frame within which more detailed analyses can be located.

Table 1.3 shows that, so far, six firms have dominated automobile production in Europe - VW-Audi, GM, PSA, Fiat, Ford and Renault - but that their relative positions changed substantially between 1982 and 1992. Table 1.4 lists the world's 20 leading automobile producers in 1990 and indicates the extent of their production both within their headquarters country and within Europe. Eleven of the world's 20 leading automobile manufacturers operated production facilities in Europe in 1990 with the new plants of the Japanese companies, Honda and Toyota, still to come on stream at that time. However, it is clear from Table 1.4 that the relative importance of Europe as a production base varies enormously between individual firms.

Three major groups of firms can be identified:

- the indigenous European producers: VW-Audi, PSA, Fiat, Renault, Mercedes-Benz, BMW-Rover, Volvo.

Table 1.3 Shares of the Western European automobile market, 1982 and 1992

	Percent of Western European sales 1982	1992
VW-Audi	12.0	17.5
GM	9.5	12.4
Peugeot-Citroen	12.4	12.2
Fiat	12.4	11.9
Ford	12.3	11.3
Renault	14.5	10.5
Japanese firms	9.3	12.0
Nissan	n.a.	3.2
Toyota	n.a.	2.6
Mazda	n.a.	2.1
Mitsubishi	n.a.	1.4
Honda	n.a.	1.3
Others	17.6	12.0

Source: Financial Times 30 October 1992; 16 June 1993

- the United States producers: GM and Ford, which have been producing automobiles from production bases in Europe for more than 70 years.
- the Japanese producers: Nissan, which began production in the UK in 1986, Toyota, which is just beginning car production in the UK and Honda, which entered Europe initially through a joint venture with the state-owned company BL in the late 1970s and has now started to manufacture at a site in Swindon. Europe not only occupies a different position in the global strategies of these three groups of automobile producers but also there are some considerable differences between firms within the same group.

Table 1.4 reveals that 8 of the world's top 20 automobile producers are indigenous European firms. Four of them - VW-Audi, PSA, Fiat and Renault - are in the top ten. However, as the final two columns of Table 1.4 indicate, they are, at least in terms of production, overwhelmingly European in their operations. Of the four firms, PSA has no manufacturing plants outside Europe. The non-European operations of the others are all confined to Latin America. By far the most important of these operations are those of VW-Audi in Brazil, Mexico and Argentina which, together, account for 15 per cent of VW's global production. VW's abortive North American operations were closed in 1988, the role of its Pennsylvania plant being taken over by VW's Mexican plant. This, in turn should be a vital springboard to serve the entire North American market if and when the NAFTA comes into existence. Fiat's only non-European production is at its Brazilian plant, which accounts for 9 per cent of the company's total production. Renault's Argentinian plant is very insignificant: a mere 1.5 per cent of Renault's global output. In fact, Renault has retrenched

Table 1.4 The importance of Western Europe as a production base for leading automobile manufacturers, 1990

Rank	Company	HQ	Total sales	Per cent in home country	Per cent in W. Europe
1	GM	USA	5208221	52.9	32.2
2	Toyota	Japan	3799921	88.1	a
3	Ford	USA	3703646	37.2	42.4
4	VW-Audi	Germany	2873869	67.2	84.8
5	Peugeot-Citroen	France	2459139	80.4	100.0
6	Nissan	Japan	2349165	86.0	3.2a
7	Fiat	Italy	1805449	91.0	91.0
8	Honda	Japan	1764775	69.3	a
9	Renault	France	1666434	79.0	98.5
10	Mazda	Japan	1302462	85.8	-
11	Mitsubishi	Japan	870112	95.8	-
12	Chrysler	USA	859245	84.6	-
13	Vaz	Russia	736050	100.0	-
14	Hyundai	S.Korea	585092	95.3	-
15	Daimler-Benz	Germany	574191	100.0	100.0
16	Suzuki	Japan	511832	100.0	-
17	BMW	Germany	499823	100.0	100.0
18	Rover	UK	464612	100.0	100.0
19	Volvo	Sweden	377878	65.8	97.9
20	Daihatsu	Japan	373110	100.0	-

a Both Toyota and Honda began production in the UK in 1992 and each expect to produce 100,000 cars per year by 1995. In that year, Nissan plans to produce 300,000.
Source: MVMA 1992

substantially its non-European operations in recent years. It closed its Mexico plant in 1986 and sold its holding in American Motors Corporation to Chrysler in 1987.

It is hardly surprising to find that Europe occupies such a dominant position in the global strategies of the indigenous European automobile producers although the extent of their relative geographical parochialism is perhaps surprising. The United States producers, GM and Ford, stand in stark contrast to the European firms. They are the paradigmatic transnational automobile producers, by far the most transnationally integrated in their strategies and operations. Not only that, but they have had such an orientation for many years, particularly within Europe itself. In fact, GM and Ford have each now evolved both a global and a European strategy but they have reached that position through different routes and over a different timescale (Dicken, 1992a). The important point from the perspective of the present discussion, however, is the relative importance of Europe as a whole to each company's strategies. As Table 1.4 shows, 32.2 per cent of GM's global production and 42.4 per cent of

Ford's global production was located in Europe in 1990. Indeed, Ford produces more cars in Europe than it does in its home country. In addition, 10 per cent of Ford's production and 6 per cent of GM's is located in Latin America and Australia. Thus, although the relative importance of individual countries to both firms varies a good deal (a point reflected also in their European operations) the overall country profile of Ford and GM is virtually identical.

In both cases, however, it is Europe which occupies the dominant position. Indeed, although both GM and Ford have been manufacturing cars in Europe for almost eighty years, the European market has never been as important to them as it has become today. In recent years, it has only been the high profitability of their European operations which has offset gigantic losses in the United States. In 1989, for example, GM's European operations generated around 44 per cent of GM's *total* net income. By 1991, Europe had become even more important to GM's financial health. In that year, GM Europe generated a $1.9 billion profit which was almost exactly equal to the company's global loss of $1.99 billion. GM's United States operations *lost* $4.6 billion in 1991 (Financial Times, 16 February 1991; 30 October 1992). GM and Ford have been engaged in a fierce competitive struggle within Europe which is clearly their major foreign battleground. Until very recently, Ford was clearly winning. In 1982, Ford had 12.3 per cent of the European automobile market and GM 9.5 per cent. However, by 1992, GM, with a market share of 12.4 per cent had overtaken Ford (11.3 per cent). Indeed, whereas GM's European operations are now highly profitable, Ford has moved from large European profits to heavy losses. In 1988, Ford's European operations generated a profit of $1.6 billion; in 1991 they generated a record loss of $961 million (Financial Times, 30 October 1992).

For virtually the entire history of European automobile production, the market for, and production of, automobiles has been shared between the indigenous European producers and the United States transnational companies (including Chrysler until it divested all its European operations in the late 1970s). In the space of a mere twenty years, however, the entire European - and global - automobile industry has been transformed by the growth and spread of the Japanese automobile manufacturers. As noted earlier, Europe absorbs rather less than one-third of Japanese automobile exports, a substantially smaller share than North America where all the major Japanese automobile producers have manufacturing operations, Japanese firms have been far more hesitant in setting up facilities in Europe. But that situation is now changing dramatically as Chapter 4 demonstrates in detail. The fact remains, however, that only three of the six leading Japanese firms listed in Table 1.4 have established car production facilities in Europe.

In production terms, Japanese firms remain strongly domestic in their orientation, although statements made by their senior executives suggest a global strategy of establishing a direct presence in each of the global 'triad' regions; a strategy of 'global localization' which, at least so far is not being followed by either European or United States producers. More than any other group of firms, however, Japanese automobile producers face two serious - and related - problems. One problem is that of market access. The spectacular success of Japanese vehicle manufacturers in penetrating national markets in Europe and North America through exports from their domestic production base has stimulated countervailing measures in the form of various kinds of restrictive arrangements (Chapter 2 gives details of the European case). The second problem facing Japanese automobile manufacturers in their global strategies arises from the exchange rate pressures generated by the huge appreciation of the yen. The kind of global strategy adopted by Japanese firms - serving world markets through exports from a domestic production base -is coming under enormous threat because of the appreciation of the Japanese currency. Each of the Japanese automobile producers is very heavily dependent on overseas markets; for example, 67 per cent of Honda's revenues are derived from exports (Financial Times, 20 March 1993). As exports have become increasingly expensive so Japanese automobile producers are having not only to restructure their domestic operations but also to expand their overseas production. In 1993, Nissan became the first Japanese automobile manufacturer to close a domestic plant since World War Two. But, as the European case demonstrates so well, expanding overseas production is a far from simple matter; in the case of the Japanese especially it is a heavily politicised and sensitive matter. It is undoubtedly the case, however, that Europe will continue to play a very significant role in the competitive strategies of all the major automobile producers. Precisely how those strategies are worked out within Europe is the subject of the following chapters. Precisely how to begin to conceptualize the context in which such strategies evolve is the subject matter of the following section.

1.3 The Automobile Production Systems Approach

According to many observers globalisation of economic activities is a major challenge for many regions and locations but may provide opportunities for some others (Amin and Thrift, 1994). Globalization, however, is an extremely broad metaphor which encompasses a wide range of cultural, economic, political and social processes (for example, see Appadurai, 1990). It is also a contested process, however, with resistances at a variety of levels from the local to the national (which has led some to refer not to globalisation but to glocalisation: see for example van Tulder and Ruigrok, 1993). Even within the more restricted domain of the industrial, it refers to very different processes in diverse fields of the social, economic and spatial organisation of industries. How then can it be delimited more precisely in relation to a production systems approach? Assume as axiomatic that firms act and respond in the light of their own economic interests, available resources and strategies and in the context of specific forms of social and cultural embeddedness. Likewise, so do their suppliers. Focusing upon interrelationships - between the actions of firms at different stages in the 'value chain', between the firm's decision-making processes and decision-making by various organizations such as trades unions and governments which enable, limit or perhaps co-manage the firm's activities - provides a new conceptual approach which can inform empirical analysis. It can also overcome some of the perceived limitations of earlier approaches. Fragmented and partial approaches to the analysis of industrial change have increasingly come under attack recently. They are no longer regarded as capable of adequately capturing the processes which simultaneously affect different economic sectors and which are deeply rooted in social and political conditions. The concept of the 'production system' has recently been suggested as a way of providing this broader perspective.

Obviously, such an approach has some advantages since it combines different elements which, together, seem to constitute a reasonably clearly definable totality. But what is the nature of these elements and their relationships? How is the system organized, how does it operate, and how is it changing? It is generally agreed that the production system denotes a 'collection of operating units linked by technology and organisation into the manufacture of final products' (Hill, 1989, 463), a kind of 'nested production' characterized by 'technical integrity' (Walker, 1988). Whereas firms and plants ('units') can be regarded as positioned into a technologically determined production chain, the production system approach goes beyond this in recognizing what has been recently termed the 'social embeddedness' of economic actions (Grabher, 1993) or, alternatively, 'their proper structural context' (Dicken and Thrift, 1992). Undoubtedly, this is a complex approach. Changes in the production system itself complicate matters still further. Yet, this approach can be seen as a useful heuristic device which helps to produce a better understanding of the complex processes which are actually unfolding in and through the European automobile sector.

We do not pretend to conduct a sophisticated debate around the concept in this chapter. Instead, we simply deploy, as a point of departure, a production system concept which has been recently used successfully in economic geography (for example, by Dicken, 1992b). According to Bailly and Maillat (1988), establishments fulfil different roles (or activities) within the overall production process: first, transformation of materials into finished or semifinished (intermediate) products; second, distribution, which embraces those activities which make goods and services available to final consumers; third, circulation which covers all intermediary activities which relate to flows, such as those of goods, information and finance and, thus, connect together the various parts of the system; and, fourth, regulation, the ways in which the system is controlled and reproduced. This includes state regulations, (new) forms of organization between firms and (new) forms of capital-labour relations etc. (Fig. 1.2).

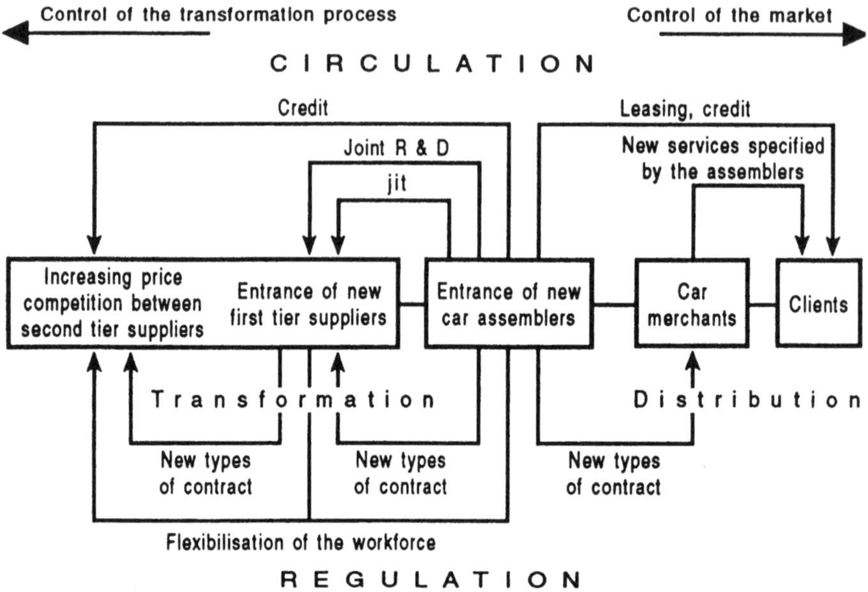

Fig. 1.2 The automobile production system

Clearly a crucial issue within such a systemic approach concerns the criteria used to define the boundaries of the system. This is, necessarily, defined by final consumption or final markets - in this case, therefore, automobile markets. As with other related but more restricted approaches (such as those of the production chain or the *micro-filière*), this meso-economic production systems approach incorporates consideration of the different stages in production from processing raw materials to manufacturing the final consumer good. In addition - in contrast to these other approaches - it also encompasses analysis of the circulation and regulation processes which, ultimately, guarantee the operation and successful continuation of the production process.

The production systems approach is clearly based upon a predominantly supply side conception of production, as consumer behaviour is treated as an exogenous factor. It could be argued that consumer sovereignty, markets and market changes should be considered more prominently as driving forces in explaining changes in production. For instance, new forms of consumer behaviour such as the abandonment or reduction in the use of private cars ('car sharing') have emerged in large urban agglomerations in wealthy countries while consumers expect new qualitative improvements in cars in regions where markets are relatively saturated, in contrast to regions in which there is a relatively low market penetration. Such improvements include enhanced passive security (via crash resistant bodies, ABS, airbags), greater comfort (as a result of improved suspension and noise insulation, for example), and increased environmental protection and ecological sensitivity (with the fitting of catalytic converters, or an enhanced capability for recycling). Obviously, the current *'leitbild'* of car consumption is increasingly called into question in some parts of Europe. These changes are not simply a result of consumer autonomy, however, for the large automobile producers are taking an active part in changing consumer tastes and behaviour as part of their marketing strategies. The sovereignty of consumer choice is therefore more apparent than real. Among others, this is one reason for focusing on the production side and its changes in this book.

The production system approach requires qualification in two further ways to enable it to become useful in empirical analysis of the automobile sector. The way it is used in this book is both narrower and broader as compared to its original macro-economic meaning (Bailly and Maillat, 1988). In this case, it is bound to the production of a specific product, automobiles. The approach is therefore used at a meso-level of analysis, as elsewhere in contemporary economic geography (for example, Dicken and Thrift, 1992). The macro-economic production system approach, however, was largely restricted in its application to national territories. But the integrity of European national states can no longer be taken for granted, while the social organisation of production is no longer limited to a particular national territory.

For the moment, though, let us retain the concept of national production systems. As different chapters in the book initially argue, national automobile production systems still prevail in Europe. This is also the message of the recent book by de Banville and Chanaron (1991) entitled *Towards a European Production System*. Such a national perspective is also emphasised by Lundvall (1992) who analyses innovation systems and rightly pointed out that these are social systems, with their dynamism grounded in social activities which are mainly nationally bound by common language and culture in communication. In fact, this touches on a wider debate about the extent to which (national) institutions still matter in global competition (Dankbaar, 1992). Conceptualizing the national automobile production system as a set of firms assembling cars and another set of automobile components firms, both supported in circulation and distribution and embedded into regulation processes, implicitly adopts an institutional viewpoint in which firms, the state and other institutions seem to be clearly defined. Different national automobile production systems may then be characterised by the kind of (national) competition among automobile producers - from one firm in Italy (Fiat) to two firms in France (PSA, Renault) and Sweden (Volvo, Saab) and more than 3 firms in UK, Germany and Spain; by the structure of the components suppliers sector more (Italy, France) or less dependent (UK, Germany) upon automobile producers (Chanaron, 1988); by the lack of protection and openness of markets (UK, Germany) versus a high degree of protection of national markets (Italy, France); by more (Germany, Sweden) or less (UK) consensual industrial relations (Dankbaar, 1992); by greater (France, Sweden) or smaller (Germany) rises in productivity and hence workforce reduction during the 1980s (Auer,

1991). As a consequence, Europe remains largely fragmented into relatively small markets. This fragmentation adversely affects global competitiveness since it results in smaller economies of scale and lower productivity as compared to Japanese and American competitors (Womack et al., 1990) and smaller automobile producer firms. The question explored in this book is not so much that of how all these apparently disfavoured production systems could produce so many cars in recent decades but that of how these systems have changed - and are changing - in response both to global pressures and to recent political events in Europe.

Radical changes are primarily concentrated in two distinct but related fields of the production system: first, in the transformation process, and, second, in the sphere of regulation. From an institutional perspective, the analysis of changes in the geography of the automobile production system in Europe requires investigation of the strategies of the main social actors involved in this process, and of the interaction between these. Four social actors can be identified and allocated to the different spheres in the structure of the production system: first, the major automobile assembly companies, with their strategies of competition and co-operation with one another; secondly, the automobile component producing companies and their relationships both to the assemblers and to each other; thirdly, workers and their institutions, such as trade unions, which shape their relationships with their employers and structure relationships between workers; fourthly, the state, primarily at a national level but also at supra-national (European Union) and sub-national levels, creating a regulatory framework which sets limits on actions both by firms and workers within the production system.

The *strategies* of the automobile manufacturers strongly determine changes in the spatial organisation of the production system but also reflect existing patterns of spatial differentiation within Europe. These strategies may be thought of as having two distinct dimensions relating to car manufacturers' attempts to control, firstly, the market, and, secondly, production. Controlling production involves consideration of both the technical and the social divisions of labour but is also affected by the changing strategies of trade unions and changing social norms about such issues as working conditions and consumption. Automobile producers have become increasingly concerned to improve their control of the market, in order to safeguard or increase market shares. There is some validity in the view that the automobile production system currently is shifting from a producing to a service system of car manufacturing (de Banville and Chanaron, 1991). According to such proponents of the French *micro-filière* approach, this transition is a consequence of the maturing product life cycle. At the beginning of its product life cycle, the technical conditions and qualities of production regulate a particular production system whereas, later, in its mature phase sales and marketing strategies become the prime steering mechanism of the production system (Jacquemin and Rainelli, 1984, 385).

Unquestionably the automobile as a product is at a rather late stage in its product life cycle. Historically, major automobile producers in Europe have not regarded comprehensive control of the *filière* as necessary in order to improve control of markets (for example, through participation in sales companies, although some car producers such as Volkswagen have had their own sales companies for many years). All that they have seen as being necessary is control of key sectors of the production system. Firms seek to manage the production system by creating market differentiation via the introduction of new improved features and qualities of cars, by tying the client more closely to a specific marque by offering new services to them (such as extended guarantee periods, "round the clock" service, and guaranteed prices for used cars), and by improving control of merchants and garages through introducing new types of contract. However, these changes in the operation of the car production system are not, as yet, fully understood. In particular, their effects on the spatial organisation of the system are not fully known, although there is clear evidence of a growing importance of urban areas as steering elements in the sales system. It may be that long established distribution patterns are changing, not least as Japanese producers in Europe have been anxious to guarantee their control of distribution by establishing their own networks. This, in turn, is beginning to lead to parallel changes by other producers, especially in the existing major national markets of Europe.

The main focus of attention in this volume, however, is on the control of production via new production concepts introduced by the automobile manufacturers and, hence, on changes in the transformation sphere of the car production system in its strict sense. Automobile producers have introduced new production concepts with totally new characteristics (referred to as "systemofacture" by Hoffman and Kaplinsky, 1988). As Best (1990) put it, the origins of these new ("lean production") approaches lie in a new consideration of time in the production process by Japanese automobile manufacturers, and in particular by Toyota. The worldwide diffusion of such new practices in the organisation of the production process both within the plant and between plants and firms has been called 'Japanization', but this process of radical change can occur without any Japanese foreign direct investment (Wilkinson, Morris and Oliver, 1992; Rawlinson and Wells, 1993). This suggest a need for extreme caution in adopting naive "culturalist" explanations of this form of production organization. It is, moreover, clear that the adoption of Japanese management practices needs adaption to 'local' circumstances. As a result, different kinds of production concepts emerge as a consequence of the firm's resources, strategies, and environment. The recent and current emergence of new production concepts can be understood primarily as the outcome of a long learning process by individual firms, which requires interactive learning within firms by management and workforces, as well as between firms by automobile manufacturers and their suppliers.

Social scientists and other analysts have mainly focussed their attention on the introduction of the just-in-time principle (and often, in the case of geographers, on the alleged association - even necessary causal link - between JIT and spatial agglomeration). There are, however, other equally - or more - important changes in the new "lean production" approaches, including the growth of vertical disintegration or quasi-disintegration, improvements in modular construction of components, and transfer of R&D to component suppliers, as well as new forms of industrial relations and work practices. New production concepts apply both to the social division of labour between firms, the technical division of labour within them and to the spatial divisions of labour within and between them. It is important to stress, however, that scale economies remain of vital importance in the emergent models of flexible lean production. They improve microeconomic efficiency in production but increase social costs such as those of transport, stress on workers (see Okamura and Kawahito, 1990), and produce problematic changes in local labour markets.

As noted above, automobile producers are not necessarily forced to extend control over the entire spectrum of the production system in order to introduce these new production concepts. Rather, a number of first tier suppliers assembling components or even technical sub-systems such as seats, dashboards and 'frontends' have been created. Consequently, direct control by the automobile assemblers can be restricted to these first tier suppliers and these restructured relationships have been the focus of recent studies (Hoffman and Kaplinsky, 1988; de Banville and Chanaron, 1991, 1993). It is important, however, that changes in the social and spatial divisions of labour as they affect second and third tier suppliers are not overlooked.

Control is predicated upon power and the asymmetrical power relations between companies within the automobile production system are important in shaping the trajectory of the restructuring of that system. Some years ago, Taylor and Thrift (1982) proposed a concept of the segmented economy, which was largely based on the relationships between large and small firms and on aggressive versus 'satisfying' strategies. Although in many European countries this large versus small firm dichotomy remained valid until recently - or, in the words of Marceau and Jureidini (1992), giants were opposing dwarfs - something new happened at the end of the 1980s and beginning of the 1990s. Even prior to this, however, some smaller component companies were not wholly dependent on the automobile manufacturers. In some countries, in particular West Germany and the UK, large component suppliers had already emerged. In recent years, simultaneously with the introduction of new production concepts in automobile manufacturing (and the extension of production capacities), there has been a considerable concentration of production in those component suppliers which could afford high expenditures for R&D. Major first tier component companies in Europe, the products of this process, are now significant multinationals in their own right (de Banville and Chanaron, 1993; Sadler and Amin, this volume). New conglomerates with considerable capital are also entering this market. It seems plausible to suggest that power relations between car producers and first tier suppliers will change further. One of the results of this could be increasing use of 'voice' in the customer-supplier relationship instead of 'exit' (Helper, 1993); that is, if problems arise the customer no longer abandons the relationship but tries to alter its characteristics. As Helper emphasizes this implies establishing a communication system between both and the emergence of relations of trust between them which, once established, enables interactive learning between customer and supplier.

Furthermore, automobile manufacturers are likely to follow at least two different strategies in their relationships to component suppliers, as has been recently suggested by Sabel, Kern and Herrigel (1991). They expect automobile producers to develop into a kind of design and marketing agency, mainly occupied with coordinating the production process and organising distribution. They contrast the model of a 'systems integrator', such as BMW, which concentrates on key capabilities and functions but seeks to safeguard technical know how via cooperation with specific independent component suppliers, with that of a 'specialization group', such as GM, where component parts production is concentrated in a particular sub-group, the GM Automotive Components Group. This ACG, however, is committed to behave in the same way as independent competitors in the supply sector.

The implications of concentration and internationalisation of component enterprises, and of the varying strategies of different automobile manufacturers, in relation to spatial divisions of labour remain, as yet, opaque. Nevertheless, it does seem clear that they will involve a reworking of production strategies in a European space, establishing large component plants to reap scale economies, rather than a recreation of regional economies around spatially agglomerated networks of component producers.

Obviously, these changes in the organisation of the transformation sub-system were not possible in the old, very sophisticated but rigid institutional *régime* which had emerged during the late Fordist era. This is not to deny the great efficiency of Fordist regulation in the context of the particular macro-economic environments of the post-war "golden age" but they no longer are appropriate, given the competitiveness of 'Toyotism' in changed macro-political and economic conditions (Jacot, 1991; Boyer and Durand, 1993). There are numerous general statements as to the necessity of changing institutional arrangements in the sphere of regulation of the production system. In Europe, where very different production models have emerged over time, there is a scarcity of reliable and valid empirical knowledge about varying institutional arrangements, as well as about their efficiency, an assessment of their future changes, and their contribution to the enhanced global competitiveness of automobile production. Most observers identify three different kinds of institutional arenas which could facilitate the more or less easy adoption of new production concepts but which have developed quite differently in different European states: first, those of industrial relations; second, the educational system, with particular reference to vocational training; and, third, those of government industrial policies. There is a rich literature about the diversified character of industrial relations in Europe, ranging from the highly conflictual (UK) to the very consensual (Sweden, Germany) (see Dankbaar, 1992; Leborgne and Lipietz, 1988). Although the conflictual/ consensual dichotomy in part depends upon the managerial culture of the automobile manufacturers, state legislation is also a very important influence. The bargaining power of unions in influencing the form of future production concepts is largely based upon laws which constitute and define the independence of unions from management (Turner, 1991). Furthermore, it is widely accepted that practical implementation of the 'flexible' new production concepts crucially requires agreements with, and the cooperation of, the workforce.

Secondly, there are further state policies which favour the implementation of new production concepts in a broader sense, in particular improvements in physical infrastructures to ease the flows of goods (such as highways) and information (such as telecommunication), technological transfer and ease of market access (Boyer and Durand, 1993). The third field of institutional differences affecting the implementation of new production concepts is, as many have stressed, in education and vocational training.

These institutional arrangements remain, so far, largely national in character. But the new map of European automobile manufacturing will not only result from the restructuring of - say - national arrangements for deregulation and re-regulation but also from these possibly becoming European ones. Given the social context of institutional arrangements, there are also important variations at both 'local' and 'regional' levels. For instance, to break traditional, somewhat rigid forms of industrial relations which are not adaptable to new production concepts within the core regions of the territory of one national state, automobile producers have relocated production to unindustrialized regions or, in the German case, to eastern regions that are undergoing a profound transition from socialist to capitalist (dis-) order. Other examples are provided by the new Fiat plant in Melfi, established so that a completely new form of work organisation could be implemented (Mehl, 1993; Conti and Enrietti, this volume), and by the Opel plant at Eisenach and the VW plant at Mosel (Schamp, this volume). Dankbaar (1992) is right to point out that 'the simple and none the less important point (is) that "location matters"'.

In the contemporary phase of radical change in the automobile production systems, however, the characteristics of an emergent institutional order which would enable the establishment of an appropriate and stable long-term context for the efficient implementation of the new production concepts remain unclear. Two different processes can be identified, which raise different questions: first, how far will the power of the national state to set the legal framework for institutional arrangements be eroded by emerging 'European' policies developed from the European Commission (Sadler, this volume) ? Certainly the national state will remain a significant actor for many years in Europe, given the diversity of institutional arrangements enabling, or restricting, the implementation of new production concepts. It would therefore be premature to herald the death of the national state (which suggests powerful counter-tendencies against, and limits to, processes of globalization of the automobile production system). In consequence, increasing contradictions are likely to emerge in the space created between the erosion of national institutional systems via the European Union (and the new opportunities that will arise in eastern Europe for car makers) and the persistence of regional and national institutional arrangements, such as those concerning industrial relations or market access. Secondly, as has been emphasized by Lung and Mair (1993), changes in institutional arrangements are the result of a complex learning process among different actors. As has been exemplified with just-in-time principles, risk aversion in the early stages of a new production concept necessitates close contact between customer and suppliers and, hence, spatial proximity. In contrast, in the later stages new institutional arrangements have been introduced, making spatial proximity redundant while safeguarding organisational proximity. In fact, what has been called, by French scholars (Lung,1994; see also Hudson, this volume), the 'hybridisation' of originally Japanese production concepts points to the way in which combined forms of old and new social action shape the emergence of new production concepts in Europe.

The incorporation of European automobile production into the global context, as outlined earlier, does not signify that Europe's production systems will necessarily become like the American or even the Japanese production system in the future. Historically evolved structures of the different parts of the production systems in Europe, as well as the strategies of the actors - whether firms or trades unions or national states or regions or localities within their territories - are too different to permit a unified system to emerge in the near future. Furthermore, some observers have demonstrated that the strategies of Japanese automakers and component suppliers in Europe differ totally from those in the USA (for example, Rawlinson and Wells, 1993). Markets remain fragmented, partly because consumers prefer buying automobiles from producers of their own country (which does not prevent Ford and Opel being seen as 'national producers' in Germany); the organisation of production will continue to differ for as long as both trades unions' strategies and national educational systems differ; locational systems will continue to be influenced - for the foreseeable future - by the policies of national states (and sub-national groups within them) to improve the attractiveness of their territories. But Europeanising and globalizing firms will further exploit and shape these differences in the course of implementing new production concepts. In this process of gradual dissolution of national production systems and the emergence of a European automobile production system, a process which is full of tensions and conflicts, definite structures have yet to become clear. All the more important, then that there is a careful empirical recording of current trends in the spatial restructuring of the automobile production system in Europe. That is one of the primary aims of this volume.

Chapter 2
National and International Regulatory Framework : The Politics of European Automobile Production and Trade

David Sadler
Department of Geography, University of Durham, United Kingdom

2.1 Introduction

The interaction between global flows of trade and investment in the automobile industry, and a vast range of national state policies, is deeply significant (see Dicken, 1992b; Sadler, 1992a). Whilst some recent programmatic accounts of restructuring in the sector (for instance Hoffman and Kaplinsky, 1988; Womack et al., 1990) have tended to downplay the significance of such measures, their influence is nonetheless real enough. Moreover, given the centrality of auto production to many individual national economies, the competitive encounter between particular companies is readily translated into friction between different states, with implications not just for the industry but also for the political context within which it operates. This chapter focuses, therefore, upon some of the ways in which distinctive interests have been articulated in the development of national and international regulatory environments for auto production in Europe.

There are numerous indications of the extent to which global change in the auto industry is dependent upon, and in turn helps to shape, political regulation. The rise of Japanese "transplant" production in North America from the early 1980s onwards, for instance - arguably the single biggest strategic shift of the decade - owed much to USA government-sanctioned trade restrictions. In that sense the evolution of production was heavily influenced by negotiation between American and Japanese governments. Such developments also embodied the shifting coalitions of those groups represented in the construction of policy. Voluntary export restraint was the product of heavy lobbying within the USA by both auto companies and trades unions; the two parties uniting in a strategic move to forge a common agenda. In this way, global change involved not just relations between states and companies, but also tangible alliances between different - often less co-operative - parties, over particular issues and in precisely - defined circumstances.

The ability of multi-national auto companies to influence (if not quite shape) the direction of certain state policies - and more broadly, to structure their regulatory environment - should not, however, be under-estimated. A large part of their power and profit-making potential rests in a global capacity in the face of national diversity. Within Europe, for instance, there are many differences in national regulatory environment despite the existence of the European Community. There is a very real complexity to the interplay of competition between companies, contrasting policy frameworks and differing trade union organisations; and global auto companies possess some very sophisticated negotiating skills. "National" interest can be defined in a variety of ways in such a context : American-owned multinationals such as Ford and General Motors have paraded themselves as good "European" citizens, for instance, united with other European producers against the Japanese challenge. This is clearly only a *tactical* alliance - and one which poses questions of national trades union structures within Europe - but it points forcefully to the web of *political* processes involved in the European auto industry.

It would be impossible within the space available here to attempt a detailed inventory of the various interest groups in Europe (companies, national and local governments, trade associations, unions and so on) and of their positions on particular issues. Instead, this chapter focuses upon two key aspects to the changing regulatory environment of the 1980s and 1990s. It first considers changes to the *national* agenda during that period, identifying the ways in which that became increasingly associated with international issues. Then it goes on to explore the development of an *international* regulatory framework under the aegis of the European Community, examining the reasons for and limits to this kind of policy response. The chapter concludes by drawing out the different implications of these interlocking environments for the key interests involved - auto companies, governments and auto industry workers.

2.2 National Regulatory Environments in Europe

Whilst it is undoubtedly meaningful (for some purposes at least) to conceive of the European auto industry as one of three major centres of world production (as it is frequently portrayed), there are other senses in which this generalisation conceals as much as it reveals. A tremendous *diversity* of national regulatory environments is one of the key factors which distinguishes auto production in Europe from that in Japan and North America. This has been significant in global terms : it was one of the reasons why Japanese assemblers directed direct foreign investment to North America before Europe, for instance. It has also been important *within* Europe, with implications for the international competitiveness of "indigenous" companies, and for the relationships between European producers and European states. More recently, these connections have become ever more tangled, with national states increasingly embroiled - either directly or indirectly - in negotiations not just with particular companies, but also with *other* European states. This section therefore describes the extent to which Europe is still far from a uniform terrain for the assembly companies, and considers some of the implications of this. It then focuses upon the ways in which relationships between national states intensified in significance during the 1980s and into the 1990s, as the internationalisation of production within the continent (see Dicken et al., this volume) became more firmly established.

The rationale for state intervention has gradually shifted since 1945. In the immediate post-war years, reorganisation and consolidation was the order of the day, particularly in the case of companies such as Renault, Volkswagen, Alfa-Romeo and SEAT. Then in the 1950s and 1960s, as national economies prospered, governments switched instead to the use of regional policy measures as a means of encouraging auto companies to locate new production facilities in areas of relatively high unemployment. From the mid-1970s onwards, as global recession and oil price rises created difficult trading conditions, a further round of crisis intervention took place. This was clearly expressed in the creation of the publicly-owned British Leyland in 1975, in the Dutch government's assistance for Volvo Car BV, and more broadly. Over the period 1977 to 1987, nearly 30 billion ECU was disbursed from national funds to individual vehicle assembly companies (see Table 2.1). This total was far from evenly distributed : one half was received by just four, heavily loss-making, "national champions" - British Leyland, Renault, Alfa-Romeo and SEAT (all these companies were publicly-owned concerns for at least some, and in several cases all, of these years). Some other major companies also received substantial support, however, including Fiat, Volvo, Mercedes-Benz, Volkswagen, Saab, Peugeot and General Motors. The extent of this assistance was one particularly marked illustration of the impact of national policies. Without such support, it is highly unlikely that the four most unprofitable companies could have survived in anything like their present form - if at all.

From the mid-1980s onwards, state aid to the sector was subject to more stringent control imposed by the European Community (see below), and in several countries there was a tendency away from fiscal assistance and towards privatisation. Individual projects meeting particular conditions still attracted substantial subsidy, however. The Portuguese government and the EC, for instance, agreed to provide 30 per cent of the 2.6 billion ECU investment costs of the joint Ford/VW plant at Setúbal expected to come on-stream in 1994 - the

Table 2.1 West European motor vehicle manufacturers : Profit (loss) and estimated state aid, 1977-87 (million ECU)

	Profit (loss)	State aid
Ford Europe	5,262	655
Mercedes-Benz	5,235	2,022
Fiat	4,986	3,212
Volvo	1,808	2,033
Volkswagen	1,540	1,563
BMW	1,133	703
Peugeot	996	1,138
Saab	969	1,512
Jaguar	432	31
MAN	270	161
Porsche	230	93
DAF	65	303
Volvo Car BV	1	430
GM Europe	(367)	1,103
SEAT	(1,705)	3,774
Alfa-Romeo	(2,305)	3,487
Renault	(4,058)	4,494
British Leyland	(4,383)	2,951
Total	10,109	29,665

Source : Bhaskar, 1990, 57

largest single foreign investment ever made in the country - in the expectation that it would provide a significant boost to the local and national economy (see Ferrão and Vale, this volume; for an earlier illustration of the bargaining process involved in such packages, see Sinclair and Walker, 1982). Moreover, some leading figures in the European manufacturers - for instance, Renault's chairman Raymond Lévy - argued that the competitive challenge from Japanese companies during the 1990s was likely to be such as to require further massive financial assistance, possibly of the order of 10 billion ECU (Financial Times, 27 June 1991)[1]. State aid will therefore probably continue to be a significant element in the European auto industry, at least for the foreseeable future.

[1] Lévy subsequently argued that his remarks had been misinterpreted:

> I did not say that the companies should be subsidised to facilitate restructuring. What I did say was that the industry needs restructuring and that this means employment problems, and that these should be taken care of by government or EC measures. (quoted in Financial Times, 22 July 1991).

However subtle the distinction, it could not hide the fact that government assistance - in one form of another - was being sought.

The varied historical incidence of such financial support is generally indicative of a broader pattern. Within some European states - principally France, Italy and Spain, but also Sweden - motor vehicle production had long been an integral part of national industrial policy. In France, both Renault (nationalised in 1946) and Peugeot were deeply-symbolic flagships of French manufacturing. Peugeot's chairman from 1984, Jacques Calvet, epitomised the intertwined relationship between car manufacturer and country in this comment, built on a re-working of the American motto that "what was good for General Motors was good for the USA" :

> I consider that what is good for France is good for Peugeot - I reverse the slogan. I cannot see how the French car industry can fight for interests contrary to French interests. (quoted in Financial Times, 21 May 1990)

Volvo's chairman Pehr Gyllenhammar expressed a similar appreciation of his company's significance to the national economy and to Swedish society, as follows:

> It is important for my country that Volvo remains Swedish. If we announced tomorrow a giant deal which was important for employment and the company's expansion, but which would mean the company having to move abroad to country x, then it would be no exaggeration to say that Sweden would suffer an identity crisis. (quoted in Financial Times, 24 February 1990)

Fiat's role in Italy was similarly wide-ranging : overwhelmingly dominant in the domestic auto market and involved in a broad range of other activities, the company proudly proclaimed in the early 1990s that it accounted for 4 per cent of Italy's Gross Domestic Product, 10 per cent of its total research and development spending and one-third of private sector R and D. It was on the twin bases of absolute economic importance, and very real social and political significance, that these companies were seen as leading elements in a national industrial policy.

Such statements as those of Calvet and Gyllenhammar - and their ready association of *national* and *corporate* interest - demonstrate the strength of the ties which fashioned government policies in at least some European states. They are also indicative of the domestic political influence of leading auto industry figures - such as Fiat's chairman Giovanni Agnelli or Peugeot president Jacques Calvet (who indicated in 1992 that he would consider standing for the French presidency in 1995). They also, however, oversimplify a complex reality. Within Europe it is possible to identify those states with a greater propensity actively to intervene, but the picture has been subject to change, both generally and in particular circumstances. For example Renault was strongly supported (at least initially) by the 1981-86 French Socialist administration, but in 1990 its status was changed from state-owned *régie* protected from bankruptcy (as an EC-imposed condition for further state aid) in an attempt to loosen its association with state patronage. In the UK, the bulk of the domestically-owned auto industry was privatised with the sale of Rover (as British Leyland was re-named) to British Aerospace in 1988. The UK government's policy stance to Rover was in stark contrast to its enthusiastic assistance to Japanese companies such as Nissan, Toyota and Honda (see Hudson, this volume). In other words the precise policies adopted at any one time in the 1980s depended upon a complex and shifting balance of conflicting interests and strategic alliances, reflecting both domestic and international developments.

Crucially, however, the *inter*-national context became increasingly significant to the direction of such policies, in both absolute and relative terms, throughout that decade and into the 1990s. One major reason for this was the deepening influence of Japanese trade and production within Europe. This expansion - important in its own right - also opened up numerous fault lines between different European states' attitudes. France and Italy, for instance, were particularly opposed to Japanese exports and imposed rigid quotas at around 3 per cent and 1 per cent of their total respective markets. This was a further re-statement of their traditionally protective attitude towards national producers (see for instance Reich, 1989). Three other EC members, however - Portugal, Spain and the UK - also operated systems of import control upon Japanese vehicles, so that together these five countries represented relatively closed markets for Japanese auto companies (see Table 2.2)[2]. The situation was made more complex by the growth of Japanese transplant production from the mid-1980s onwards, which was heavily concentrated in the UK. The UK's government's attitude to Japanese assemblers rested upon their potential transformatory impact upon labour relations, automotive component manufacturers, and other assemblers (see Hudson, 1992a). Other European states perceived Japanese-badged production in the UK as an additional and unwelcome challenge in an already-competitive marketplace.

In the late 1980s several countries - principally France and Italy - sought to exclude Japanese vehicles produced in the UK from their domestic markets, on the grounds that they were not sufficiently "European" in origin. The precise definition of "local content" was a particularly contentious point at issue in this regard. The UK government's ex-works price basis - whilst relatively easily monitored - begged several questions to do with the extent to which its

---

[2]These national restrictions existed because EC countries were allowed to keep import controls they had in place (that were acceptable to GATT) before joining the Community. Hence Italy imposed a quota on Japanese cars in the 1950s in retaliation for import limits upon the sale of Italian cars in Japan. Spain and Portugal also had quotas for the sale of Japanese cars, which were established before joining the EC. The French government had no such legal recourse but relied instead upon exploitation of the differing technical standards for cars sold within member states of the EC (hence its refusal, during the 1970s, to sanction the last stages of EC-wide technical harmonisation). The UK had no formal control over Japanese imports but depended instead upon an informal agreement established in the mid-1970s. The range and diversity of bilateral trade restrictions was but one measure of the barriers to a truly common single internal market within the Community.

Table 2.2 Japanese penetration of European Community automobile market, 1987

	Total sales 1000s	Japanese sales 1000s	Japanese sales as a per cent of the total market
Ireland	54	24	44
Greece	51	18	35
Denmark	124	40	33
Netherlands	556	144	26
Belgium	406	84	21
West Germany	2,916	441	15
Luxembourg	30	5	15
UK	2,014	225	11
Portugal	129	9	7
France	2,106	63	3
Italy	1,977	14	1
Spain	852	2	-
Total	11,213	1,069	10

Source : Financial Times, 17 November 1988

inclusion of indirect costs and profit margins enabled assemblers still to source technologically crucial high value-added items - such as transmissions - from Japan. Fiat claimed in 1988 to have dismantled a UK-built Nissan Bluebird piece-by-piece, identifying only 21 per cent of the direct production cost as unambiguously of European origin (with 46 per cent Japanese content and 33 per cent of unidentifiable nationality). At that time, the UK government and Nissan maintained that the car had over 60 per cent local content. The ultimate target for Japanese assemblers in the UK - one agreed with the government - was to reach 80 per cent local content, and Nissan achieved this in 1990 with the launch of its Primera range. Even that was not enough for some, however, such as Peugeot's chairman Jacques Calvet, who argued that local content should be nearer 100 per cent than 80 per cent. Such issues were important in a trade dispute which simmered (without ever really coming to the boil) for much of the late 1980s and early 1990s.

These questions of trade and production were particularly significant within the context of the EC, and their (partial) resolution could only come from within that forum (on which, see below). In other words the Japanese challenge could only be handled *politically* through a broad *international* framework within Europe, albeit one which encompassed many wide-ranging differences of attitude. A further aspect to the deepening internationalisation of political regulation of the auto industry in Europe, however, came through the combination of state involvement in particular companies, and a growing tendency towards corporate strategic alliances and/or mergers. This was particularly clearly evident in the tangled relationships which gradually emerged from the late 1980s onwards, binding Renault, Volvo, Mitsubishi and the French, Swedish and Dutch states.

In 1990 Volvo and Renault unveiled the terms of a technical and industrial co-operation agreement, backed by an exchange of equity. Volvo took an initial 20 per cent stake in the Renault parent company (with an option to increase this to 25 per cent within three years) and a 45 per cent share in RVI, Renault's bus and truck operation. In turn, Renault took a 45 per cent stake in Volvo's commercial vehicles business and a 25 per cent share of its car operations. The deal created the world's biggest commercial vehicle manufacturer, and was intended to be backed up by further co-operation in product development and component purchasing for both cars and commercial vehicles. The package was a major extension of existing co-operation between the two companies. Renault already supplied engines and transmission components to Volvo Car BV in the Netherlands (and held a minority stake in Volvo for the first half of the 1980s in an earlier, unsuccessful attempt at collaboration). For both companies it was an attempt to secure their future via further internationalisation. It fell short of a full-scale merger, though, mainly because of fears over potentially adverse *national* reactions. Even the limited share-swap agreement required the formal approval of both French and Swedish governments.

The following year this relationship became still more complex. Volvo Car BV - in which the Dutch government held a 70 per cent stake - agreed terms with Mitsubishi for a joint venture agreement, giving the Japanese company its first auto production base in Europe. The Dutch state was to sell a 33.3 per cent stake in the company to Mitsubishi, and 3.3 per cent to Volvo, creating a structure of three equal shareholdings. The deal could only proceed, however, with the tacit support of the French government which - via its 90 per cent stake in Renault, now linked to Volvo - had previously expressed strong opposition to further expansion of Japanese production in Europe. In this way an axis of power emerged linking the French, Dutch and Swedish states in a particular form of response to corporate restructuring, one which increasingly posed questions over the reality of a "national" interest, even in the strongholds of such views[3]. As the subsequent collapse of the Renault-Volvo merger

[3]Such a tangled grouping was far from universally welcomed. By mid-1992 Renault and Volvo bought around 15 per cent of their automobile components from common suppliers, and intended to lift that proportion rapidly to around 80 per cent, also establishing a joint quality control venture. Yet a French government-commissioned report, sharply critical of Volvo's production management and marketing efforts, was 'leaked' by a French union official, Gerard Muteau. His accompanying comments - that "there is a real danger in pursuing this accord" (quoted in Financial Times, 9 July 1992) - reflected French union concerns, particularly over Volvo's alliance with Mitsubishi, and emphasised the dilemmas which internationalisation posed for certain interest groups.

demonstrated, however, questions of "national interest" remained significant in a global auto industry (see Malmberg, this volume).

This combination of linkages was but one illustration of the enhanced significance of relations between states by the early 1990s, as corporate alliances in the auto industry intensified in both frequency and complexity. In part, such co-operation was driven by differing reactions to a major "external" challenge - Japanese investment within Europe -but it also hinged upon an intensification of internationalisation processes within the continent, expressed both economically and politically. A further element in this equation came with the opening-up of eastern Europe as a potential market and production platform, following dramatic strides towards liberalisation there from 1989 onwards. This is not the place to document the expansion of automotive investment in eastern Europe in detail (see instead Swain, 1992), but it is important to note its political implications, both there and in western Europe.

Before 1989, national states had been heavily involved in the central co-ordination of auto production within eastern Europe, as a particular form of division of labour emerged, one that was heavily conditioned by each member of the Soviet bloc's relationship to the USSR. The Hungarian company Ikarus, for instance, specialised in the production of buses and coaches, whilst the Czech company Skoda (amongst others) produced passenger cars. After that date, western auto companies - especially VW, GM and Fiat -moved rapidly to take over out-dated production facilities and to invest in new factories, particularly in the former East Germany, Hungary, Czechoslovakia and Poland. This both led to and was encouraged by a new and different form of state intervention within these countries (see Kapitány and Kállay, 1991).

For some companies - particularly VW - expansion eastwards was part of a strategy wherein newly-acquired facilities, principally in eastern Germany and Czechoslovakia, were to be integrated into a broader European car production system as rapidly as possible (see Schamp, this volume). VW's chairman Carl Hahn commented also that his company's move into eastern Germany gave it the opportunity "to fulfil a role that is not only *industrial* but *political* too" (quoted in Financial Times, 13 March 1990, emphasis added). This far-from-coded reference to the re-unification process was a clear statement of the sub-agenda to western automotive investment in eastern Europe. Many east European states heavily supported the attraction of such capital flows. In 1991, for instance, the Czech government granted VW (whose purchase of Skoda represented the biggest single privatisation and foreign investment deal in the region to that date) a substantial tax "holiday", and protection in the domestic market from the import of both new and used cars. The privatisation process also re-defined the nature of state activity in eastern Europe. Mlada Boleslav - home to Skoda - was formerly dependent upon the state-controlled company not just for employment, but also for a whole range of social provision, such as education. Following its purchase of the company, VW transferred these functions to the local and national public authorities, divesting itself of certain responsibilities and recasting the boundaries of state activity within the country.

Such transformations were important not just for the companies concerned but also for the states (and the implications of eastwards expansion were significant too, but in different ways, for those companies - such as Peugeot and Renault - which failed to make major investments, at least in the first wave). They indicated some of the ways in which national state regulatory environments had shifted in response to growing external challenge (in the east) and to the threat of lower labour cost competition (in the west). Employees and managers of SEAT, for instance, were reported in 1990 as expressing alarm about the implications of VW's investment in the east for its plans in Spain. Whilst labour costs were only one - relatively minor - element in the overall economics of production, VW, Mercedes Benz and other German-based auto assemblers and components manufacturers planned to cut back sharply on employment in western Germany in the early 1990s, citing high labour costs as a primary cause (see Hudson, this volume, Table 4.1). Erika Emmerich, president of the German auto industry association VDA, referred to this disadvantage as follows:

> However we look at it, our production costs are too high. The wage costs of the German auto industry are almost a third higher than the Japanese, they are two-thirds higher than the British and the French, and 50 per cent higher than the Spanish. High wage costs and - by international comparison - short working hours significantly reduce the competitiveness of the German industry. (quoted in Financial Times, 20 February 1992)

In these remarks lay an important clue to one of the attractions of eastern Europe (from the standpoint of western capital) and an equally significant indication of the deepening problems for workers in auto companies.

The increasingly - internationalised European car production system, and the prevalent national regulatory mechanisms, together spelled out a looming dilemma, if not yet crisis, for these employees. Put simply, the origin of this problem lay in the limited extent to which trades union representation in the industry was organised on an international footing. The first European-wide regulatory structure in the auto industry was agreed by VW and its unions only in 1992, covering at first only the company's operations in Germany, Spain and Belgium (although it was intended subsequently to extend the remit to embrace Czechoslovakia). The situation at Ford was more typical. There, negotiations on the establishment of a European works council were delayed by major differences between competing national interpretations. British trade unions wanted full-time officials to be members of such a body; the German union IG Metall preferred instead direct works-based elections of Ford employees. In Spain, Belgium and France trades union representation was deeply divided on religious and political grounds, whilst in the company's other European plant (newly-established in Hungary) independent trades unionism was only just being re-established. The contrasting environments were also reflected in some very real differences in working conditions : Ford employees at plants in Cologne and Saarlouis worked a 37-hour week in 1992, whilst those at Dagenham and Halewood put in a 39 hour week. Some of the difficulties of European co-operation were neatly encapsulated in this comment from Wilfrid Kuckelkorn, IG Metall chairman of the Saarlouis plant's works council:

> We want the British unions to win a 35-hour week [due to be implemented in Germany by 1995] and they will get practical solidarity from Germany including overtime bans and working to rule [But] the national unions cannot accept European negotiations. If you take away the power of the national unions, they will go down. (quoted in The Guardian, 25 October 1991)

In a variety of ways, then, the national regulatory environment of auto production within Europe shifted decisively during the 1980s and early 1990s. National state aid continued to be significant for some investment projects, but more generally the pattern of public expenditure in the sector was increasingly tightly controlled, partly by the EC. State activity began to shift more and more towards an internationally-oriented focus, either in terms of relations with Japanese producers or of the complex dealings between different European companies. The paradoxes of such intervention were at their sharpest in the enhanced interaction of "national champions". The agenda was also altered by political liberalisation in eastern Europe, posing new problems for states there (of adaptation and privatisation) and in the west (of enhanced competition in the provision of attractive preconditions for profitable production). Together, such changes further facilitated capital's ability to seek the best possible deal on its own terms, not only from governments but also from employees. The intensified internationalisation of production and the gradual re-focusing of national state activity created very real problems for trades union organisations, particularly as they remained grounded in (often sharply antagonistic) national bargaining structures.

Given this emergent emphasis on international regulation of the industry, the following section goes on to explore in more detail the ways in which European Community policies towards the sector developed during the 1980s and early 1990s. It describes the points of connection, and the increasingly evident tensions between this regulatory framework and that provided by its constituent member states.

2.3 International Regulation: the European Community's Policies Towards Auto Production and Trade

From the mid-1980s onwards, the European Community's own policies towards auto production and trade increased in both sophistication and significance. In part this was driven by the same processes which affected individual national states' policies, particularly the internationalisation of production and the competitive challenge from Japan. The EC also had its own, ideologically-laden agenda, epitomised in the "1992" programme and grounded in a neo-liberal approach to industrial policy. This frequently took the EC into deep conflict with some of its own, more readily-interventionist member states. This section therefore first describes the motivations for and objectives of EC policies, and then considers some of their implications (for fuller details, see Sadler, 1992b).

From the outset, the EC strongly emphasised the benefits which might be expected to flow from a single internal market in automobiles. One volume of its fifteen-volume research project into the "costs of non-Europe" focused on the automobile sector and argued that "1992" would have a dynamic impact through liberating trade, encouraging parts standardisation, lowering costs and increasing demand (see Ludvigsen Associates, 1988). The Cecchini report (1988) took automobile production as a prime example of the potential gains from a single internal market, summarising the benefits which could be reaped as 2.6 billion ECU, or 5 per cent of the industry's total costs. Such arguments, however, underplayed the extent to which existing auto companies had already begun to treat Europe as a single production base, even though national markets remained very fragmented (see CEC, 1992a; Dicken, 1992a). They also ignored the potential impact of Japanese producers, which were subject to a wide range of restrictive practices. These included member states' differing import controls within Europe, their contrasting attitudes towards Japanese transplant production, and a range of other issues to do with technical harmonisation as a precondition for freedom of trade. In other words the ethos of the "1992" programme actively necessitated a round of EC policy intervention to remove the internal barriers to free trade, and to resolve - or at least patch over - the differences between member states.

A further element of the EC's "1992" programme - the creation of fair competition - involved the monitoring and gradual reduction of national state assistance to industry. A Community framework on state aid to the motor vehicle industry was established in 1989 (see Official Journal of the European Communities, 1989). Its overall philosophy was expressed as follows: "Over-reliance on state aid to solve problems of industrial adjustment *vis-á-vis* third country producers undermines competitiveness of the Community car manufacturers by hindering the economically healthy influence of market forces". The framework's objectives were therefore "to establish full transparency of aid flows to the industry and impose at the same time a stricter discipline to the granting of aid". It required advance notification to the EC of aid from each member state on any project with a total cost exceeding 12 million ECU, and an annual report from each country detailing how the subsidy was disbursed. The framework specified that only state support either for restructuring linked to a specific and satisfactory plan, or in the form of regional policy incentives, or for vocational training, or pre-competitive research and development, would receive a favourable hearing. Support for rationalisation, as part of normal company operations or to cover operating costs, was unlikely to receive approval.

This declaration confirmed the Commission's disposition against national state support to vehicle manufacturers, except in precisely - defined circumstances. It was a policy area where the Commission's powers could effectively be enforced, by recourse to legal action if necessary. In 1988, for instance, the Commission had obliged the UK government to cut from £800 million to £469 million the subsidy paid to British Aerospace on the occasion of its purchase of the former state-owned car company Rover. In 1990, the Commission and the French government agreed a compromise deal under which Renault paid back half of a Fr 12 billion subsidy it had received in 1988, previously approved by the Commission subject to certain conditions (particularly a reduction in capacity) which had not been properly met.

The issues of trade control and transplant production were to prove, however, much more challenging. An accord negotiated in the early 1980s between the EC and Japan's Ministry of International Trade and Industry (MITI) - after Japanese imports had grown from 1 per cent to 9 per cent of the EC market in the 1970s - established an informal 'monitoring' process on Japanese sales to Europe. In 1987 these amounted to 1.1 million vehicles, or around 10 per cent of the market (see Table 2.2). A target of 1.2 million vehicles was agreed for 1988. So sharply divided were the competing member states' attitudes, however - and so significant were the questions at stake - that a long and drawn-out negotiating process was to ensue. It was not until 1989 that Industry Commissioner Martin Bangemann secured agreement, even within the EC, over a working document on a single Community motor vehicle market, let alone with Japan (see CEC, 1989a). This represented a compromise between the free-market views of the Commission and some governments, and the protectionist stance of other countries, notably France and Italy. It defined the task as follows : "the Commission is determined to ensure, for a clearly defined period after 1992... that Japanese vehicle exports do not grow to such an extent that they risk creating problems within the Community" (p.13).

On transplant production, it was if anything even less precise : "account must be taken of this form of production in defining future solutions, bearing in mind that the Community is fundamentally in favour of direct foreign investment and is anxious to encourage better integration of such production into its economy, while abiding by its international commitments and without resorting to compulsory local content formulas" (p.14). This deliberately vague position signalled that a lot of hard negotiating still remained to be done.

Much of this, of course, rested upon the credibility (or otherwise) of the Commission's forecasts. These were presented in a companion document, investigating the future of the motor industry (CEC 1989b). This contained some valiant assumptions (for details see Sadler, 1992b). It was founded on the premise that demand within Western Europe would increase by 1 per cent annually over the period 1988-1998, from 12.9 million to 14.2 million, and that production there would grow from 13.0 million to 14.5 million. Of the latter total, it was assumed that Japanese transplant capacity would amount to 1.2

million vehicles annually. This expansion alone - well below some other expectations - would prove a considerable challenge to established European manufacturers. Within a global picture, it was estimated that Japanese exports to Europe would decline, as Japanese domestic production fed instead into developing country markets - although the exact reason for this predicted switch was not made explicit. In other words if transplant production expanded *alongside* continued Japanese exports, then competition in Europe for sales was likely to be even more intense.

Given the fragility of the relatively optimistic scenario upon which the Commission based its preliminary findings, it was perhaps not surprising that establishing consensus on these important topics continued to be a formidable challenge. Negotiations continued throughout 1990 over the precise stance which the EC should take in talks with Japan, and became increasingly rancorous. Divisions between member states were mirrored by conflicts between vehicle producers. Jacques Calvet, president of Peugeot, was particularly intransigent in his opposition to Japanese exports from plants in the UK:

> The UK will have to decide in favour of its national interests or for Europe. If it chooses Europe....the Nissan, Toyota and Honda plants will enter into the overall limit of Japanese cars. (quoted in Financial Times, 3 February 1990)

Suggestions that transplant production might be included amongst any transitional arrangement drew a response from the companies concerned. Yutaka Kume, president of Nissan, commented for instance that:

> We are doing our utmost to increase local content levels in our plants. Is it fair to regard these cars produced in Europe as the same as cars exported from Japan? (quoted in Financial Times, 21 June 1990)

The UK government argued strongly in favour of the interests of such Japanese producers located in the UK (see for instance European Communities Committee, 1990). Not all European manufacturers took quite the same line as Peugeot, although most agreed on the need for some form of protection for a period after 1993. Robert Eaton, president of General Motors Europe (in a neat twist, allying his company with "indigenous" European producers) preferred to emphasise the need for "closing whatever efficiency gaps might exist *vis-á-vis* the Japanese competition" (quoted in Financial Times, 5 October 1990). The situation was worsened too by the ending of the late 1980s sales boom, as market conditions deteriorated drastically. In this environment tensions grew close to breaking-point.

Agreement between the EC and Japan was finally concluded only in 1991, after lengthy negotiating sessions which mirrored the time taken to establish an outline package. The deal had a shadowy backcloth, for formal limits on transplants would be questionable under existing EC law, although the Japanese parties were anxious not to allow mounting tensions to escalate into a full-scale trade war. *Informally*, therefore, a seven-year transitional period was adopted in which the number of Japanese vehicles sold in the EC would be limited to about 2.4 million by 1999, around 16 per cent of the estimated European market (and note that this assumed market, of 15.1 million in 1999, was higher than that previously adopted by the EC - 14.2 million in 1998 - partly because the deal also embraced light commercial and off-road vehicles). After 1999, there would be no constraints on the sale of Japanese vehicles in Europe. In the interim, there were to be no restrictions on the sale of vehicles produced by Japanese companies within the EC, although there was an expectation (on the EC's part at least) that such transactions would not exceed 1.2 million units annually.

Almost immediately, however, there were several indications that not all parties were happy with the outcome. Peugeot's president Jacques Calvet regarded it as a "detestable and extraordinarily ambiguous agreement" (quoted in The Guardian, 2 August 1991). His fears highlighted some of the grey areas of the proposal : its lack of formal status, the limited room for manoeuvre by the EC if Japanese market share expanded more rapidly than anticipated, and the assumption that transplant capacity would be restricted to 1.2 million vehicles annually at a time when other estimates suggested even stronger growth, to possibly 2 million autos annually. Such fears were fuelled by an alternative interpretation of the ceiling of 1.2 million vehicles produced by Japanese companies inside the EC, offered by Nissan's president Yutaka Kume (also president of the Japan Automobile Manufacturers' Association). This limit, he said, was:

> not a figure we have agreed between Japan and the EC. This is the EC's unilateral internal assumption. How much is produced [by Japanese companies] in Europe will be decided by how the European market develops. (quoted in Financial Times, 12 September 1991)

President of Renault (and of the European car manufacturers' association ACEA) Raymond Levy responded that members of the Community had to avoid a temptation to compete with each other for further Japanese investment : "all EC states should act together, and not run after the Japanese and beg them to set up factories", he argued (quoted in Financial Times, 18 September 1991).

The EC's "consensus" with Japan, then, appeared only superficially to have papered over the cracks of internal dissent, let alone establish a secure framework for international trade. It was subject to the differing interpretations of member states over what represented an appropriate industrial policy, and the EC's neo-liberal line - in this as in other sectors - was sharply discordant with that in at least some European capitals (see also Sadler, 1992c). Long-awaited Commission proposals for a motor vehicle policy, finally published in 1992, envisaged enhanced support for pre-competitive research and development in the sector, but drew short of specific training and re-training measures. The initiative was left firmly in the hands of European companies, on all counts : "the responsibility for such changes lies first and foremost with the industry itself" (CEC, 1992b, 19). As the EC increasingly attempted to constrain national states' roles in guiding industrial restructuring (one part of its free-market ideological stance), however, it left European industry more and more open to global competition. At the same time its policies towards auto trade were the product of intensely secretive negotiations, within the Commission as well as with Japan - a point not lost upon the European Parliament (see for instance European Parliament, 1991). Yet the difficulties encountered in reaching agreement on even a limited range of issues also pointed forcefully to the problems in the path of any broader EC-wide auto industry policy framework. In this, the *international* regulatory environment contained as many, if not more, difficult dilemmas as did the national arena.

Such a situation prompted a number of responses. One view was that the rest of the 1990s would see further dramatic restructuring within Europe, particularly in consequence of the expansion of Japanese-owned facilities: Pemberton (1991) for instance argued that the number of major indigenous companies was likely to fall from six to five by the end of the decade. The impact of Japanese direct foreign investment took on, from the nationalist perspective, almost cataclysmic dimensions. Lehmann (1992, 50) suggested that the most likely scenario for the 1990s was one in which "the French automotive industry disappears, relegated to industrial archaeology or - at best - French automotive factories become subcontractors and distributors of Japanese and other foreign multinationals". Whilst the language employed might differ, similar fears - whether right or wrong - lay behind the tenacity and diversity of the European response.

An alternative reaction to the EC's increasingly evident problem in dealing systematically with major global issues and power structures was recourse to the local, and in particular to champion the position of local and regional government. For example, Elliott and Gray (1991, 262) argued as follows:

> The dominant influence of national governments defending national interests has proved an inadequate basis for Europe to respond to worldwide trends in the vehicle industry in the 1980s : it is likely to become increasingly inadequate in the 1990s. Mechanisms need to be found to allow local and regional interests to transcend those of national governments and contribute towards European policies.

Their favoured device included not just individual local authorities but also broader national and international groupings (see also Davenport et al., 1990). The real power of such bodies was of course debatable, especially in relation to national governments; it must also be remembered that there was a very fine distinction between mutual support, and the competitive promotion of particular places and regions as bases - both actual and potential - for auto industry investment. From the splintered diversity of interest groups fragmented around the European regulatory system it was important not to lose sight of the key capacities of major actors, and the ways in which they differed.

2.4 Concluding Comments

One conclusion is inescapable : from the 1980s onwards, the increasingly international framework within which auto production and trade took place posed an abundance of new challenges for the European regulatory system. For national governments - in both east and west - these involved issues to do with intensification of competitive pressures, a growing prevalence of international corporate alliances, and the limits - both fiscal and practical - on their capacity to intervene. There was an increasingly evident structural imbalance in the relationship between automobile manufacturer and national government. This was partly why a growing emphasis was placed upon regulation by the European Community, as one means of addressing the questions of Japanese exports and transplant production in particular. The tensions affecting even this limited exercise in international co-operation, however, were to prove decisive limitations on the EC's capacity for action. In all of this, the other major interest group - the employees of auto companies - experienced a period of, by and large, great uncertainty. Its position was rendered even more insecure *vis-á-vis* the auto companies than that of national governments (and in part this was a result of deliberate governmental policy choices). The impediments - real and imagined - to Europe-wide trade union co-operation in the face of corporate globalisation were a revealing insight into the new power structures of the European regulatory environment in the 1990s.

Chapter 3
"Europeanisation" in the Automotive Components Sector and Its Implications for State and Locality

David Sadler[1] and Ash Amin[2]
[1] Department of Geography, University of Durham, United Kingdom
[2] Centre for Urban and Regional Development Studies, University of Newcastle upon Tyne, United Kingdom

3.1 Introduction

The absolute and relative significance of the automotive components sector became increasingly apparent during the 1980s. Assembly companies in North America and Western Europe successfully sought to out-source greater proportions of the value of a finished vehicle (following the highly profitable Japanese model), and to delegate enhanced responsibilities to component companies in the field of design and development. These moves were associated with an intensification of restructuring processes, leading to the emergence of a "first tier" of major companies producing automotive components (although it is important to note that they were not necessarily geared exclusively to production for the auto industry), and to heightened relations of dependence within the supply chain between these companies and their own suppliers. As the assembly industry responded to intensified global competition, in other words, further changes were initiated both to the relationship between assembly and component activities, and within the automotive component sector.

This chapter is primarily concerned with the implications of change in the component industry, although this cannot completely be isolated from developments within the assembly sector. It seeks to draw out some of the preconditions for and consequences of internationalisation in the manufacture and supply of automotive components within Europe during the 1980s and early 1990s. During those years, wholesale shifts began to take place in terms of the geography of the component industry. These had dramatic implications both for national state policies and for particular localities where such activity either was, or might be, or had been, concentrated.

The argument proceeds as follows. It starts in one European country - the UK - using this as a window on broader processes. The decline of Rover (formerly known as British Leyland) and of the West Midlands as a base for automotive component manufacture in the early 1990s is contrasted with resurgence of the UK-based assembly industry since the late 1980s (largely as a result of Japanese investment) and its implications for patterns of component production and supply. National state policies were deeply implicated in such changes, in terms not only of the financial regime at Rover (state-owned from 1975 to 1988) and the support offered to Japanese companies (particularly Nissan), but also of the climate underpinning new labour relations agreements. The shifting use of the UK as a base by automotive component companies over these years is also established as an introduction to the concept of "Europeanisation" within the supply chain, which forms a basis for discussion in the following section. Detailed accounts of production strategies in the component sector within Europe are used to distinguish between the differing responses of European, American and Japanese owned companies to change in the European market, and to identify the implications for state and locality. These issues are then addressed further in the concluding remarks.

3.2 The Changing Geography of Automotive Components Production in the United Kingdom

The historical dominance within the UK automobile market of those companies that combined to form British Leyland in 1968 was reflected in the emergence of a classic "industrial district" of small and medium sized engineering companies centred in the West Midlands around the major UK-owned car assembly plants. There, the production of goods for and the supply of services to the auto industry was heavily concentrated, so much so that the fortunes of the UK motor industry acted as a kind of barometer for the economic prospects of the West Midlands region. During the 1970s and for much of the 1980s, however, this appeared to point in just one direction : deepening depression. In the face of mounting competition, BL's share of the UK market (where its sales were heavily concentrated) fell in the 1970s from 40 per cent to 20 per cent, and even to below 15 per cent in the 1980s. Drastic measures were taken in an attempt to stem mounting losses. In a five-year period at the depth of recession during the late 1970s and early 1980s, capacity was cut by more than one-third and the workforce slashed by over a half. These moves reverberated throughout the West Midlands, as the knock-on effects of declining orders and reduced purchasing power exacerbated the problems of an already-fragile local economy. In this way the pressures of international competition (particularly the expansion of Japanese and other imports, and of sales by other companies making vehicles in the UK such as Ford and GM), mediated by a particular form of state policies, were felt forcefully in a region where automotive component manufacture had historically been concentrated.

These changes also had broader implications for the UK automotive components industry. During the 1970s this was heavily dependent upon sales to motor manufacturers in the UK; one survey of firms in the West Midlands in 1976, for instance, found that two-thirds of them made no export sales whatsoever. Despite this, Rhys (1979) found grounds for optimism about prospects for expansion into European markets by UK companies, particularly by take-over. Events were, however, to take a very different course. As UK-owned car production capacity contracted, component firms which had long been tied to a narrow domestic market found instead that other, non-UK companies were making massive inroads via sales to and (increasingly) direct foreign investment in production for those assembly companies which were taking up BL's market share. Particularly in the recessionary early 1980s (when raising capital for fresh investment was rendered prohibitively expensive by the UK government's interest rate policy), many UK component firms were forced to contract dramatically or even quit the sector altogether. Increasingly too control passed via take-over not, as had once been anticipated, *by* UK companies but rather *of* such firms by foreign investment. In 1977 43 of the country's 60 biggest component companies were domestically-owned; by 1988 that figure had fallen to 28 (see Sleigh, 1988).

Changing fortunes within the assembly industry therefore impacted very directly upon the UK component sector, especially in the crucial period of the late 1970s and early 1980s when other European component firms were initiating and accelerating their own processes of internationalisation. By the late 1980s it appeared that only a handful of UK-based companies were of sufficient size and financial strength to contend with the price-squeeze being exerted by assemblers, and to meet the demands for investment in production and design facilities associated with "single-source" supply contracts (see Amin and Smith, 1990). These firms had followed very similar strategies of tightly focusing upon particular specialist engineering areas (transmissions at GKN, engine management and braking systems at Lucas, engine components at T & N, friction materials and brake and clutch components at BBA, narrow tubing at TI), and disinvestment from the production of lower margin commodities (steel manufacture by GKN, exhausts by TI, electrical equipment by Lucas). Those components which remained in production at these companies formed only part of a portfolio of other activities (such as aerospace at Lucas) which gave additional ability to weather cyclical trends in the auto sector (and some of these companies - especially Lucas, BBA and TI - intended to shift still more business into these alternative areas, as a medium-term strategic objective). It was no coincidence that these companies had begun to internationalise even in the 1960s, well before the onset of crisis in the UK-owned vehicle industry in the 1970s, giving them a broader base and greater scope for accelerated internationalisation and product diversification in the 1980s.

By and large, though, these companies were the exception. Some one-time major actors in the sector made strategic decisions to cease automotive component production altogether: Smiths Industries, for instance, sold its vehicle instrumentation interests to Lucas (which subsequently closed them down) and focused instead upon the aerospace industry. Some such as Associated Engineering and Automotive Products were taken over by other UK companies (T & N and BBA respectively). Many more either closed their doors for the last time or were taken over by non-UK companies, as they found themselves choked off from international expansion by a combination of domestic government policies and falling demand. This changing pattern of ownership was ultimately reflected in the European rankings of major component manufacturers : only one UK-owned company, GKN, figured amongst the top ten by sales in 1990 (see Table 3.1).

Internationalisation of production was of course a multi-faceted phenomenon. For those UK companies which did penetrate international markets via exports and/or direct foreign investment, there was the prospect of greater flexibility in production and cost-reduction through economies of scale (although this did not necessarily strengthen the job security of those working at factories operated in the UK). Moreover, as assembly companies increasingly sought to enter into "partnership" arrangements with leading component companies in certain product ranges across national and even continental boundaries, only those firms with a broad international presence looked likely to participate in the design, development and volume production of new generation components during the 1990s. The same held true for other European and foreign-owned companies investing in the UK as a base for the manufacturer of components. Yet crucial functions such as control of investment decision-making, and research and development activity, tended to rest within a firm's parent country. In this way, the limited presence of UK companies among Europe's leading suppliers posed questions not just in the short-term, but also - and perhaps eventually more significantly - for the longer-term.

These issues were given particular importance by the wave of Japanese investment in vehicle assembly which was heavily concentrated in the UK from the mid-1980s onwards (see Hudson, 1992a and this volume). Not only were these companies seen as being at the forefront in developing new forms of relationship with the component sector; they were also highly significant in terms of their impact upon existing spatial patterns of component production and supply. For some (such as Hoffman and Kaplinsky, 1988), drawing on

Table 3.1 West European automotive components industry, 1990

Company	Sales $ billion	Country of parent
Bosch	10.8	Germany
Michelin	10.3	France
Philips	4.8	Netherlands
Continental	4.6	Germany
Valeo	3.9	France
Magneti Marelli	3.4	Italy (Fiat)
ZF	3.1	Germany
BASF	3.1	Germany
GKN	2.8	UK
Pirelli	2.6	Italy
Lucas	2.5	UK
ACG-GM	2.4	USA (GM)
Goodyear	2.2	USA
Teves	1.8	USA (ITT)
BTR	1.8	UK
Allied-Signal	1.8	USA
TRW	1.8	USA
Tenneco Automotive	1.7	USA
Saint-Gobain	1.6	France
T & N	1.5	UK
Pilkington	1.3	UK
Mahle	1.2	Germany
SKF	1.2	Sweden
VDO	1.2	Germany [1]
BBA	1.2	UK

Source : Sleigh 1991, 16-17

[1]Purchased by Mannesmann in 1991

experience in the USA (where some 300 Japanese components facilities were established close to new Japanese assembly plants during the 1980s), there was a necessary and logical association between just-in-time production and spatial proximity of supplier to assembler, giving a new twist to old ideas of industrial location theory. In the West Midlands, at least, Nissan's choice of Sunderland in north east England for the first Japanese assembly plant in Western Europe was initially portrayed as a threat to existing component manufacturers (see for instance WMIDA, 1989). The establishment of such new plants, it was argued, would jeopardise the prospects of other assembly companies and of their long-established suppliers. From early indications, however (Nissan only produced more than 100,000 cars per year in 1991, and Toyota and Honda only commenced production at their plants at Burnaston and Swindon respectively in 1992), it appeared that such fears were over-stated or, at least, over-simplified (see Amin and Smith, 1991). Japanese assemblers initially placed great emphasis on working with existing component companies in Europe, rather than encouraging further rapid internationalisation on the part of the Japanese components industry (see Sadler, 1994). There were a number of reasons for this : scale differences between the American and UK contexts; strong political pressure from European governments anxious to encourage rapid development of high "local content"; and the limited absolute extent of the potential market for the Japanese in Europe, at least in the early 1990s, to name but a few.

Nonetheless, the presence of Japanese auto assemblers in the UK did raise new questions, of state policies towards foreign investment and trade (see Sadler, this volume) and of locational choices - at a variety of scales. For some products - particularly those with frequent variation in product specification, such as seats and seating mechanisms - there was a rationale in close proximity to the assembly operation. Others could be supplied over longer distances via frequent exchange of order information and the use of external warehousing. For some (including electrical systems such as starters and alternators), economies of scale meant that one or a handful of sites could supply the whole European market, making national selection rather than location within any given country of primary significance. As the UK car industry recovered from the mid-1980s onwards, and as manufacturing costs in the UK tumbled, there was a fresh incentive for such facilities to be located there - though (and crucially) not necessarily by UK-owned companies. In different ways, then, the UK automotive components industry was being re-shaped in the late 1980s and early 1990s, as new and varied forms of "localisation" accompanied the continuing globalisation of the automotive industry.

It would be premature to offer a definitive account of the result of such processes. Some informed speculation is possible, however, not least about the intentions of Japanese assemblers. Nissan's choice of north east England probably reflected its central concern for the installation of new labour relations practices and attitudes in a region with no prior tradition of vehicle assembly, for instance. Toyota historically paid greater attention to the development of a locally-clustered supply complex in Japan, and it was perhaps no coincidence that it located nearer to the old strategic heartland of the UK components industry in the West Midlands, and to emerging centres of such activity such as south Wales. Honda's partnership with Rover seemed to give it the best of two worlds : access to selected West Midlands suppliers on its terms, and the opportunity of developing new arrangements in product areas where Rover's original equipment suppliers failed to meet the desired standard. For all three companies, there was a very real prospect of integration with and modification to existing patterns of component production and supply.

Part of the outcome will also be determined by trends inside the UK component industry. Nissan, for instance, rated the quality of its UK supply base in the early 1990s at 65-70 per cent of best Japanese levels, slightly below that of its continental European suppliers, although UK companies were improving at faster than the standard European rate (see NEDO, 1991). The cost-advantages of manufacturing in the UK (particularly with respect to labour) and a resurgent UK assembly industry were significant locational attractions for other European companies : the market leader Bosch opened its first-ever UK components plant in 1991, for instance. In this respect, the influence of history ran deep, for a major element of the restructuring that was taking place involved such greenfield investment and/or take-over by other European companies from outside the UK. Despite improved competitiveness (achieved partly through working with Japanese customers such as Nissan) the UK components industry still had a long way to go, and was participating in the process of "Europeanisation" in a distinctive fashion, one that was heavily conditioned by its highly-destructive experience in the early 1980s. Even in the context of a growing market opportunity offered by Japanese assembly investment, in other words, an earlier round of decimation and limited internationalisation meant that the bulk of the UK-owned component industry (or at least what was left of it), as opposed to companies making components in the UK, still suffered some severe handicaps.

In these ways, then, the UK automotive components industry became increasingly integrated into a broader European production system during the course of the 1980s and early 1990s. This process of "Europeanisation" also took a variety of different forms in other countries, but effectively it signified a breaking-down of the barriers between national systems. Leading European companies emerged which were capable of organising production not just across the continent, but increasingly on a global scale, emulating if not quite matching the degree of internationalisation prevalent in the assembly industry. The UK was a distinctive participant in this process for a variety of reasons, particularly the impact of national government policies in the late 1970s and early 1980s, and the wave of Japanese assembly company investment from the mid-1980s onwards. There, re-casting of the assembly sector via different forms of globalisation - especially by Japanese capital - was associated with tentative signs of changing forms of localisation, as the component industry was drastically re-organised (though not necessarily by either Japanese or British capital). Similar concerns, however - if in different ways - also occupied assembly and component companies and national governments in many other European countries. In Germany, for instance, relatively high production costs in the late 1980s and early 1990s led to a search by both major assembler and leading component companies for increasingly decentralised patterns of component supply, linking with the UK and other low-cost regions such as southern Europe (see Schamp, this volume). Integration, in other words - and its manifestation in the form of flows of investment and trade - carried implications both for source and destination countries and regions. The following section therefore goes on to consider some of the differing ways in which automotive component companies participated in this "Europeanisation" process, and to evaluate some of the implications.

3.3 "Europeanisation" and Corporate Strategies in the Automotive Component Sector

The complexity of a modern automobile is mirrored in the structure of the automotive component industry. Many different products, processes and technological skills are combined in a contemporary vehicle, and these are supplied by a similarly broad sweep of companies with primary interests and expertise in many different activities : mechanical engineering, electronics, chemicals and material manufacture, to name but a few. This trend has even been intensified as vehicle assemblers seek to externalise the purchase of ever-greater proportions of components rather than produce "in-house". At the same time, there have also been on-going processes of concentration involving consolidation of the position of a small elite group of major suppliers; and of centralisation, involving expansion into the sector by companies previously regarded as specialists in other fields of activity. Whilst the overall picture is a complex one, therefore, it is possible to draw a number of general conclusions about the evolving structure of the West European automotive components industry.

Estimates of the absolute number of companies involved in producing automotive parts and materials in one form or another vary substantially. A Boston Consulting Group/PRS (1990) study commissioned by the European Community estimated that there were 3,250 such firms in the EC, but Sleigh (1991) regarded a figure of even 1,500 as unrealistically high, for instance. There is probably a large and shifting "tail" of very small enterprises, particularly in some countries. On the other hand, the identity of the leading 25 or so companies is clearly evident (see Table 3.1). They include some diverse groupings. In the field of electronic components and systems, for example, the Dutch giant Philips had a very strong presence alongside its German rival Siemens (in 26th place by sales in 1990) and (since 1991 when it purchased the instrumentation company VDO to add to other newly-acquired assets in the sector) the German "steel" firm Mannesmann, whilst the leading components maker Bosch also specialised in the application of electronics to mechanical systems (although to an extent this connection was also prevalent in other engineering companies). The UK specialist Pilkington maintained a close rivalry with the French engineering conglomerate Saint-Gobain for the supply of glass. Four companies specialised (though not exclusively) in the production of tyres : Michelin, Continental, Pirelli and Goodyear, along with the diversified UK-based BTR group, which also had substantial engineering

Table 3.2 Sales and employment at selected automotive component companies, 1991

	Total world sales		Automotive component		World
	Own currency (billion)	US dollars (billion)	as per cent of total	US dollars (billion)	employment 1000s
Bosch	DM 33.6	22.1	55[2]	12.2	181
Valeo	Fr 19.7	3.8	100	3.8	27
Lucas[1]	£2.5	4.7	59	2.8	55
GKN	£1.9	3.6	67	2.4	31
T & N	£1.4	2.6	65	1.7	39
BBA	£1.3	2.4	48	1.2	23
Allied-Signal	-	11.8	35	4.1	98
ITT	-	20.4	14	2.9	110
Tenneco	-	13.7	12	1.6	89
TRW	-	7.9	51	4.0	71
Nippondenso		9.1 (1989)			

Source : company annual reports

[1] Year to 31 July

[2] Estimated: automotive components division represented 49 per cent of turnover but this did not include the sale of in-car entertainment systems by the mobile communications division under the Blaupunkt label.

interests in the sector . The German chemicals concern BASF vied with DuPont of the USA (in 28th place). Two one-time national champions, Valeo and Magneti Marelli, had expanded to varying degrees beyond national frontiers (although the latter was directly controlled by Fiat, rendering its true "independent" status questionable, as was the case with GM's component group). There was also a range of other engineering firms specialising in various components and systems : two from Germany (ZF and Mahle), four headquartered in the USA (ITT, Allied-Signal, TRW and Tenneco), four from the UK (GKN, Lucas, T & N, BBA) and the Swedish bearings group SKF.

To a very real extent, the strategies of these leading companies evolved in relation to those of their major actual and potential rivals in distinctive product areas. Market leadership was increasingly important, especially in terms of generating sufficient revenue to fund expensive research and development and investment programmes. Competitive advantage lay in a high market share for a specific family of technologies, processes or products. This had to be measured not just in one or a handful of national markets, but across Europe (via supply contracts established in "partnership" with major assemblers), and increasingly on a global scale. Potential competition abounded, particularly from the threat of Japanese investment but also from other countries and other firms diversifying into the sector. Enhanced internationalisation within clearly-defined product ranges was one strategic response adopted by leading companies in order to maintain their market position in Europe, both by European and American-owned companies. The account below therefore starts by considering the evolution of some of those firms, focusing on Bosch, Valeo, the group of four major UK companies and the four leading American suppliers (for comparative information on these companies see Table 3.2; see also Sadler, 1991c, 1992d). It then goes on to consider a prospect which occupied considerable attention at practically all these companies, especially from the late 1980s onwards : that of a round of Japanese investment in the component sector, matching that by assembly firms.

3.3.1 Bosch

Robert Bosch GmbH was Western Europe's (indeed the global) leading independent supplier of mechanical and electronic automotive components in the early 1990s, and a major multi-national enterprise in its own right. It produced a broad range of parts, although anti-lock braking (ABS) and engine management systems represented particular areas of specialism. Whilst still strongly focused on its original bases in Germany (particularly around Stuttgart), Bosch's automotive component activities had become increasingly internationalised, to a greater extent than other parts of the group. More than half of the workforce in this sector, and half of the total of over 50 separate plants, was based outside Germany. Major facilities were operated in 20 countries including Spain and France, the USA, developing economies such as Brazil and Malaysia, and (via a joint venture) Japan. This combination of locations reflected a desire to minimise manufacturing costs and to develop close collaborative relationships with major vehicle manufacturers wherever they were located, particularly for the component families in which it specialised. One estimate was that it had a 70 per cent world market share in fuel injection and anti-lock braking systems (see The Economist, 25 April 1992), giving it considerable bargaining strength in these areas.

For the Bosch group as a whole, Europe represented by far the dominant market area, accounting for over 80 per cent of sales (note that although this pattern was unlikely to be exactly replicated in automotive components, relevant data were not publicly available; the group's position as a GmbH under German accounting laws meant that only limited disclosure of such information was necessary). Its operations in France (its largest foreign market, closely followed by the USA) and in Spain were closely connected to an integrated European system, with over one-third of French output and over one-half of Spanish production being exported. The proportion of investment targeted outside Germany turned sharply upwards in the later 1980s, practically doubling within three years to reach two-fifths of the total. New plants were constructed and existing ones expanded not only in Europe and the USA, but also in the developing world. This trend was partly continued during the early 1990s, as falling demand and rising labour costs in Germany encouraged still further relocation of production (although it was partly offset by expansion into the new east German states : within a year of re-unification, Bosch employed 4,000 people there). During 1991 Bosch shifted production of Blaupunkt car radios and speakers to lower-wage factories in Malaysia and Portugal, reducing its German workforce in these products by over 1,000. Capital investments of DM1 billion were planned in France, Switzerland and Turkey over the period 1990-92. Additionally, a DM 320 million project involved construction of a plant in the UK, to produce a new generation of compact alternators. This was the company's largest ever single foreign investment.

Establishment of this second generation compact alternator plant at Miskin, near Cardiff, represented a significant extension to Bosch's corporate strategy, taking it into direct contact with Japanese automobile companies in the UK from its own production base there for the first time. The plant began production on schedule early in 1991 with a workforce of 560 and an annual capacity of 2.5 million units. The UK was chosen in preference to Spain as a site for the first plant making this product (although a second facility was expected to come on-stream in Spain in 1992) partly because of the resurgence of vehicle production there, led by the wave of Japanese transplants. "We came to Wales because we saw an increasing base in the UK as the motor industry expanded, though an important part of our output will be exported", said Joachim Burkhardt, Bosch's technical director in Cardiff (quoted in Financial Times, 7 January 1991). The UK, along with Spain, also offered substantial labour cost and other manufacturing advantages over Germany. Under a planned second stage, output was expected to double and the workforce to expand to 1,200 by 1995.

The Cardiff plant's product was both lighter and some 20-70 per cent more powerful than earlier models, designed to meet fuel efficiency requirements and the electrical demands of more sophisticated vehicle management systems. It went into production sourced from imported components, and management there admitted to some difficulties in finding suitable UK-based suppliers. In other words the initial operation was very much of an assembly-plant in character. It was also geared to a very wide market : up to 80 per cent of the output was to be exported, and the UK and Spanish plants were to be the only global sources of supply for this particular line. As such it was just one element in Bosch's broader corporate strategy (although rendered potentially more important by the distinctive blend of labour relations practices that was instituted). It also fitted neatly into the company's distribution strategy, which rested heavily on the use of external warehousing : completed alternators were generally shipped to stores either in the UK or Germany (for mainland Europe), to be "called off" subsequently against customer requirements. Crucially, the product had been developed in Germany and this was closely related to the initial sources of basic componentry. A substantial part - over 90 per cent - of the group's research and development spending was concentrated there, despite the internationalisation of production which had taken place. The ways in which Bosch re-evaluated the UK as a potential production platform, then, not only provided a further expression of continuing Europeanisation of the sector, but also revealed some of the developmental limitations of particular forms of inward investment.

3.3.2 Valeo

Valeo, the leading independent French mechanical and electronics components manufacturer, went through a round of massive restructuring in the late 1980s in response to operating losses made from 1984 to 1986. These deficits were largely a reflection of a slump in French car production and the financial troubles of the Renault and Peugeot groups, upon which Valeo then depended for a large proportion of its sales. The production of alternators and starters caused particular problems, after it had taken over a series of ailing French manufacturers. From 1987 onwards though, the Valeo group (under new management control) returned to profit, in a fashion that provided a sharp contrast to the UK automotive components industry's response to downturn in the market there just a few years previously. It became wholly dependent upon the automotive sector in 1988 after the last disposal of peripheral activities (which had accounted for some 10 per cent of turnover). Several leading French companies in the sector were taken over in a bid to strengthen Valeo's position there. It also initiated expansion in the geographical spread of operations, so as to be less dependent upon the fluctuating fortunes of the French vehicle assembly industry. Turnover increased rapidly, practically doubling to reach Fr. 20 billion by the end of the decade. This dramatic growth helped to secure Valeo's position and enabled it to approach a size where it could sustain massive internally-funded investment and research and development programmes during the 1990s.

The French businesses which were taken-over included Neiman (specialising in security systems and at the time the third largest French component company) and Chausson (manufacturers of heat transfer equipment), both acquired in 1987 and an electronics company, G. Cartier Systemes, bought in 1989. These represented a continuation of Valeo's attempts to secure strategic re-organisation within the French automotive component industry, efforts which had been on-going since the 1970s and were tacitly backed by the French government. A still more ambitious hostile offer for Epéda-Bertrand Faure only failed in 1988 in the face of a counter-bid from Peugeot and Michelin, companies which were anxious to restrain the emergence of this highly-powerful interest in the sector within France. Valeo's international acquisitions were even more substantial. In 1989 for instance, it purchased the Blackstone Corporation of the USA (supplier of engine cooling systems) and Delanair of the UK (a producer of air conditioning and wiping units). At the same time it invested in new facilities, opening a factory in Turkey to supply clutches to local assembly plants opened by Fiat and Renault, and establishing new operations in Mexico and Brazil. Joint ventures were also used as a route for expansion, in South Korea and the USA as well as in Spain (the latter in a plant with Nippondenso to produce ignition coils, which subsequently ceased to be part of Valeo's portfolio). Revenue earned outside France as a proportion of group sales more than doubled as a consequence of these moves, to reach 40 per cent. By 1991 there were some 65 Valeo plants in 15 countries around the world.

The key to this expansion - both domestically and internationally - was a strong focus upon particular product areas in which Valeo had a major market share. It was the world leader in clutches, clutch facings (friction materials), and engine cooling; the European market leader in climate control, lighting and security systems; and it occupied second place in the European market for alternators and starters. It was no coincidence that these seven component ranges comprised the bulk of Valeo's nine major operating divisions (the others being wiping systems and electronics). In the early 1990s the company entered upon a phase of consolidation (after several years of frantic growth), divesting itself of some activities which had become peripheral to its long-term vision, such as brake linings and some electronic components. This served further to focus Valeo's product range. Re-organisation also entailed the closure of some plants and facilities, mostly in high cost manufacturing sites such as France and Germany.

The company had become both a leading actor in the French auto industry and a major European player. Its four-point industrial strategy unveiled in 1990 incorporated a commitment to significant volumes of research and development, quality and cost improvements, training and skill development, and partnership agreements with its own suppliers. These were all indicative of a position as a substantial force in the European component sector. This status had been secured via a concerted programme of expansion and acquisition, organised around defined product ranges, a strengthened position in the domestic market, and intensified "Europeanisation". Like the market leader Bosch, Valeo's research and development activity remained firmly concentrated in the company's parent country, even as its operations spread more widely across Europe (and this focus upon one continent was clear : it represented 84 per cent of total sales in 1991). Unlike Bosch, Valeo expanded in the UK largely by takeover. Its four locations there (climate control at Ammanford, clutches at Birmingham, security systems at Blyth and wiping systems at Hengoed) owed much to the earlier acquisition of companies such as Neiman and Delanair (although the Ammanford and Birmingham plants in particular, and also that at Blyth, saw substantial new investment in the early 1990s). The end result, however, was similar : a process of "Europeanisation" controlled by a strong domestic market leader involving trans-European production strategies.

It is also instructive to contrast the attitudes of national governments to restructuring in the sector : in France and Germany (and in Italy) the emergence of one component sector giant was encouraged whilst in the UK, as shown below, it was actively discouraged. This was significant not just in strategic terms, but also as part of a commentary upon the differing ways in which patterns of component sourcing and supply were being established at a variety of scales : local, national, and European.

3.3.3 Four major UK companies : Lucas, GKN, T & N, BBA

A measure of the difference in scale between leading UK and other European component companies was that the combined turnover in the sector of the top four UK firms in 1991 was less than that of Bosch (see Table 3.2). The sales of the top two UK companies together only just bettered those of Valeo. This position was a reflection of a variety of factors. Over-dependence upon a collapsing domestic market in the early 1980s forced the UK component industry into a round of radical restructuring, involving specialisation upon selected products and the abandonment of some others. Additionally, the non-automotive interests of several firms expanded markedly, effectively further diluting their presence in the automotive sector. Whilst all four became specialists within their chosen ranges, in other words (and took substantial European market shares), the overall product spread of their automotive activities was very narrowly-defined. This held a competitive advantage in that it clearly focused corporate goals; the question mark which it carried was whether such strategies were capable of generating sufficient cash-flow in the long-term for the continued renewal of businesses via expensive research and development and investment programmes.

All four major companies, then, underwent turbulent change during the 1980s. Lucas was stimulated into action by the company's first-ever loss in over one hundred years of trading (made at the depth of recession in 1981), one that was wholly attributable to its automotive division (which also operated unprofitably in 1983). It emerged from this process far less dependent upon such activities (accounting in 1991 for 59 per cent of sales against 72 per cent in 1981), expanding instead its interests in aerospace and applied technology. GKN too changed shape markedly, shifting out of the group's original product areas in steel manufacture and forging (moving such assets into companies like Allied Steel and Wire and United Engineering Steels), and growing in the industrial services sector, with the overall effect that the production of automotive components accounted for an increased proportion of group turnover. The two smaller companies - BBA and T & N - consolidated their position in this sector by merger, acquiring Automotive Products and

Associated Engineering respectively in 1986. For both, however, the production of automotive components was still only one part of a more diversified corporate portfolio, encompassing other engineering interests. T & N made around two-thirds of its sales in the auto sector in the early 1990s, although this was largely a consequence of its desire to move out of its construction materials holdings. It had a long-term goal of balancing sales equally between automotive and other engineering applications. BBA's share of turnover derived from the auto sector tumbled from 80 per cent after its acquisition of AP to below 50 per cent by 1991, with a strategic goal of even further reduction. By and large then, these companies were far from totally dependent upon the automotive sector, and indeed most placed or intended to place decreasing *relative* emphasis on it. In this context it is also significant that a fifth firm, TI, which had built up a substantial auto-related presence in the 1980s (accounting for one-quarter of its sales in 1987) made a strategic decision not only to divest itself of certain such assets (like other companies; see below) but also to expand in other areas, a choice which was cemented by its £500 million acquisition of the aerospace group Dowty in 1992. This carried a close parallel to Smiths Industries total withdrawal from the automotive sector in the early 1980s and represented an extreme version of strategies being practised by some other automotive component firms.

These companies also undertook extensive changes *within* their automotive divisions, concentrating on a narrow range of products and divesting or ceasing to produce others with limited profitability and/or market share potential. Lucas, for instance, ended electrical component production in Birmingham in 1986, and sold off its lighting division and its starters and alternators businesses (the latter to Magneti Marelli) during 1987. It specialised instead in the production of chassis systems (braking and brake management), power train systems (centred on engine management - for instance fuel injection for diesel power units) and electronics (for example sensors and wiring products). GKN similarly stressed a narrow family of components : about 80 per cent of its automotive turnover came from the production of transmission equipment, particularly from its leading market position in the manufacture of constant velocity joints, drive shafts and universal joints. T & N concentrated upon bearings and engine components, such as pistons, particularly after it acquired GKN Vandervell in 1988 (giving it a dominant position in the UK market : an argument which had - somewhat paradoxically, perhaps - been central to the Monopolies and Mergers Commission's opposition to GKN's attempted purchase of AE [subsequently bought instead by T & N] in 1983; see Amin and Smith, 1991). BBA focused on friction materials for braking and power transmission systems (an area from which T & N was disengaging, particularly as it sought to minimise the legacy of its former asbestos - processing interests), claiming to be the leading European supplier in this broadly-defined range (although note that Valeo - market leader in clutch materials - no longer emphasised friction materials for braking systems).

At the same time, these companies all similarly responded to crisis in the UK market by expanding into other regions : not just mainland Europe, but also North America. In 1981 for instance Lucas made 46 per cent of its total sales in the UK and 75 per cent in Europe; by 1991 these proportions were 32 per cent and 70 per cent respectively, with the contribution from North America trebling to over 20 per cent. A similar picture unfolded at GKN : 59 per cent of its sales came from the UK and 81 per cent from Europe in 1981, but 34 per cent and 76 per cent respectively in 1991, with the American element doubling to 18 per cent. At that time T & N still produced 50 per cent of all its output in the UK (of which half was exported) and 80 per cent in Europe; just 10 per cent of sales were made in North America. At BBA, by contrast, the European component of its turnover declined from 85 per cent to 50 per cent in the second half of the 1980s, and the American contribution practically trebled to reach 35 per cent. Such changes were reflected in dramatic employment shifts. At Lucas, UK employment fell from three-quarters to one-half of the group's total; at GKN, employment in the rest of Europe exceeded that in the UK for the first time during 1991.

These companies - formerly geared to the UK market and (in some cases) the legacy of the British Empire - had become by the early 1990s major European actors, developing sophisticated trans-European production systems. Some of them also invested in the American market; one element in a strategy of co-operation on a global scale with leading customers, but one that carried heavy financial burdens, especially given their limited absolute size and their commitments to growth in Europe. This was one reason why several of them participated in a number of joint ventures, particularly with Japanese companies. In the UK for instance, Lucas established such concerns in the late 1980s with Sumitomo (Lucas SEI, in which Lucas held a 70 per cent stake, for manufacture of wiring systems) and with Yuasa (Lucas Yuasa, 50 per cent owned by each partner for the production of batteries). In the USA another operation - Lucas Sumitomo Brakes, based in Ohio - produced brakes. T & N also developed a joint venture with Japan Brake Products to produce friction products near Nissan's plant in Tennessee.

Such partnerships increased in frequency during the late 1980s and early 1990s, and were underpinned by a number of motivations, including the strength of Japanese technology. UK-owned companies were far from alone : many other European, American and Japanese companies participated in similar arrangements. Access to additional capital resources was a significant reason, however, particularly in the UK context. In this there lay a clue to the long-term legacy of these leading UK component groups' chosen path. Whilst product specialisation was a highly rational management strategy within certain parameters, it depended upon capturing and maintaining very high market shares to generate adequate revenues for re-investment. In this context, it was probably vital that these companies gained a substantial element of business in their respective product ranges with the Japanese assembly companies which established in the UK; although, of course, they needed to invest in long-term development programmes to hold on to such contracts. Similarly, whilst expansion into non-UK bases - both in mainland Europe and North America - was vital for any company with ambitions to become and/or remain a global force, it carried heavy financial costs. The true effectiveness of these companies' response to the crisis of the early 1980s, then, would be assessed by whether it enabled them further to pursue such activities in the 1990s, or whether the relative attractiveness (i.e profitability) of alternative business directions - combined with the competitive challenge of other European and American, and probably also Japanese, companies - would force them to reconsider their strategic objectives for and presence in the automotive components sector.

3.3.4 Four major US companies : TRW, Allied-Signal, ITT and Tenneco

The strong presence of major US automotive component companies in Europe in the 1990s frequently went unremarked, especially in the context of emergent debates about Japanese investment (see below). This was so despite the extent of their operations, which closely matched the ways in which the American vehicle companies Ford and GM had established co-ordinated production systems across the continent. Companies such as TRW, Allied-Signal, ITT and Tenneco were major actors in the European automotive components industry and, more generally, were substantially larger in overall scope than practically all European-based component companies, particularly since the American companies' European operations generally formed only one part of holdings diversified both geographically and by product range. In other words, these companies had developed sophisticated policies of production within Europe over the years which not only matched those of their European rivals there, but also drew greater strength from overall balance-sheet power.

Each of the leading US component companies established in Europe, then, produced auto parts as just one element of their operations. Allied-Signal also operated in aerospace and engineered materials, having increased turnover four-fold since its days as Allied Chemicals, via the acquisition of Bendix Corporation in 1983 and The Signal Companies in 1985. TRW participated in defence equipment and services; Tenneco in six other areas (including pipelines, shipbuilding and chemicals); and ITT in eight others (including electronic components, insurance and hotels, the latter under the 'Sheraton' name). The global automotive component businesses of Allied-Signal and TRW ranked alongside the total volume of Valeo's sales, even though they only represented one-third and one-half of parent company turnover respectively. At ITT and Tenneco, automotive components accounted for barely one-eighth of group turnover yet these divisions still ranked alongside those of Lucas and GKN. These were very large corporations, particularly by the standards of

European component companies, with a sophisticated ability both to develop in particular market areas and to generate substantial cash-flows.

These companies had a particularly strong European presence in the automotive sector, as part of a concerted policy to occupy a strategic position in two major market areas (North America and Western Europe). Allied-Signal had been active in Europe since 1924. After rapid expansion in the 1980s, it had nearly 15,000 employees in the sector there at over 40 plants, in what it claimed was the best strategic position of any US-based auto supplier in Europe. TRW made just under one-half of its sales in the sector in Europe from a similar-sized base of around 40 factories, concentrated in Germany, France and the UK. ITT made over half of its sales of automotive components in Europe particularly through its subsidiary Alfred Teves, a German-based brake manufacturer acquired in 1967. Tenneco operated some 14 plants in Europe and had begun a concerted process of further expansion there, epitomised by its acquisition of the UK company Armstrong Equipment in 1989. Others in this elite group also continued to expand in Europe; Allied-Signal, for instance, bought Valeo's friction materials business in 1990.

Within their chosen product ranges, this broad geographical spread was both a result of and condition for substantial global market shares. Allied-Signal was the world leader in two important areas : braking systems (with a 33 per cent share) and friction materials, and also in turbochargers. It was less dominant in the production of ABS, a point of some concern as this increasingly became standard rather than optional or luxury equipment, although it had begun to expand into other products, such as the production of passenger safety or restraint systems, especially airbags. Tenneco was world leader in the production of exhausts (under the name of Walker) and of suspension control components, particularly shock absorbers (under the name of Monroe). TRW focused on the production of steering and suspension parts, restraint systems (including airbags), engine valves and fasteners. ITT's Alfred Teves specialised in the manufacture not just of brakes, but more particularly ABS.

These were not just big companies, then, whether in overall terms or in the sense of their auto-related interests. They were global market leaders in their chosen product ranges by virtue of their substantial presence both in North America and Western Europe. There was a stark contrast between both their size and their scale of operations in two continents, and the leading European component companies. Typically around 50 per cent of the American companies auto parts revenue came from Europe, whereas most major European companies (with a few exceptions, such as BBA) depended on Europe for over 80 per cent of turnover, with only 10-20 per cent coming from North America. In a world where high initial development and investment costs were becoming ever-steeper, such a substantial market presence - backed by massive parent company resources - was a powerful competitive advantage. The longer the production run of new and/or proven technology, the greater the cash-flow to sustain profitability and further investment. As well as operating sophisticated Europe-wide production systems, then, such companies were on their way to true global status.

3.3.5 Japanese component companies and Europe

Much the same was frequently said of some Japanese component companies in the late 1980s and early 1990s, though probably with a greater degree of hyperbole. For in the wake of a massive expansion in production capacity by Japanese assembly companies in North America during the 1980s, there was a similarly dramatic flow of direct foreign investment there by the Japanese automotive components industry. Around 150 Japanese auto parts firms established some 300 plants in North America during the decade. Assemblers such as Nissan, Toyota and Honda largely encouraged this development, finding USA suppliers heavily-dependent upon contracts with American assembly companies. In Europe, the subsequent onset of Japanese assembly plant investment there was greeted from some quarters with fears of a further similar incursion by the Japanese component industry. Opinion was mixed on this issue, however; one which was likely to have deeply-significant long-term implications for the future of the auto parts industry in Europe during the mid to late 1990s.

The pace of Japanese component company investment in Europe was largely limited initially, and mainly indirectly, by the tensions of investment friction. Japanese assembly companies moved much more slowly and cautiously in Europe than in North America. The first major car plant, Nissan's factory at Sunderland, only produced more than 100,000 cars per year in 1991, at a time when 11 facilities in North America were capable of producing in excess of 2 million vehicles (total Japanese output in Europe at that time was of the order of 0.2 million). One survey estimated that there were only 50 Japanese components plants in Europe in 1990, of which over half were in the form of joint ventures (Boston Consulting Group/PRS, 1990). The Nissan Sunderland factory's supply base of 195 companies in 1992 included just 10 Japanese firms, with a further 12 joint ventures involving Japanese participation. At the same time, there was a view which held that the technical quality of the European auto components industry (including that owned by American companies) was superior to that which had existed in North America on the eve of Japanese assembly plant investment there, and that this was one factor likely to curtail Japanese parts companies' ambitions in Europe (see for instance Womack et al., 1990, 164; note also that companies such as Toyota and Nissan went out of their way to play down fears of Japanese investment in the components sector in Europe).

On the other hand, the Japanese component industry was no less anxious to internationalise than its European or American counterparts, particularly as its main customer base increasingly produced off-shore. Typically this involved investment not just in North America, but also in south east Asia, another favoured location of the Japanese assemblers. Moreover, as the volume of Japanese-owned assembly capacity in Europe steadily increased, so did the attractiveness of establishing a production base there. Whilst output levels in the early 1990s hardly represented a profitable market, a forecast volume of nearly 2 million by the late 1990s was a far more viable proposition, particularly if combined with the opportunity of supplying other European or American assemblers. Frequent model changes by Japanese car manufacturers - with the prospect of a new generation at the earliest-established plants of Toyota, Nissan and Honda from the mid-1990s onwards - could well provide a suitable market opening. Many of these companies also had substantial financial and technological resources, being either or both of a major presence in the sector in global terms (if nothing else by virtue of their position in the Japanese market), and part of one or other of the Japanese keiretsu, or inter-firm groupings. Even Nippondenso, for instance - clear leader by sales, twice the size of Valeo and nearly on a par with Bosch's automotive turnover - was 23 per cent owned by Toyota.

Within the UK - where Japanese assembly capacity was concentrated - there were some incipient signs of further Japanese investment in the component sector, although certainly not on a scale (at least in the early 1990s) yet to warrant comparison with the situation that evolved in North America during the 1980s. For example, the UK market for radiators was effectively captured by Japanese capital in 1989 as Rover Group's subsidiary Llanelli Radiator was sold to a Nissan affiliate, Calsonic (formerly Nihon Radiator), and Nippondenso purchased IMI Radiators for £25 million (subsequently re-named ND Marston and the recipient of a £16 million investment programme which came on-stream in 1991). In the same year Nippondenso created a joint venture in Spain with Valeo (see above, section 3.2) to produce electronic ignition components. It subsequently assumed full control of the plant (which came into production in 1991) as Valeo withdrew from this market area. In 1990 Nippondenso also announced the establishment of a £65 million project at Telford in the UK, established jointly with Magneti Marelli (though with Nippondenso holding a 75 per cent stake and management control). The plant - which began production during 1992 - made air-conditioning and heating systems, and aimed to produce 0.8 million units annually by 1995 with a workforce of some 450. This placed the company in direct competition with Valeo for this product range from its own base within Europe, and it was a formidable competitor; contracts to supply Toyota and Honda in the UK were rapidly established.

On a global scale, Nippondenso was a clear contender with other leading giants in the sector. Its product strengths were heating and air conditioning and electrical components (accounting for around one-third of sales each) and engine management systems (15 per cent of turnover). Its financial strength was built on major market shares in Japan - not just with Toyota, which took half its output there - but more generally. It had just over one half the Japanese market in radiators, air-conditioning and fuel-injection systems, and claimed world leadership in the production of alternators, starters and radiators, and second place in fuel injection systems behind Bosch (with which it had a long-standing technological-sharing agreement). It had been active in North America since the mid-1980s. By 1991 its sales in Europe amounted to £170 million, mostly from Japanese exports; it anticipated sales of £450 million there by 1995, with over one-half coming from European-based production.

Nippondenso was, of course, a far more substantial company in terms of absolute size than many of its rivals in Japan; although these were largely strengthened by association with much larger resources within their *keiretsu*. Nonetheless, Nippondenso's incursion into and intentions for the European market were quite probably symptomatic of broader developments within the Japanese component sector; or at least of its aspirations. For if market leadership rested on global sales and production, then Europe was an obvious target once bases had been established in North America and south east Asia. For these companies, this was a highly rational policy (at least in their own terms); for the components industry in Europe, it would add (and indeed already had contributed) a further layer of uncertainty. The outcome of the competitive struggle between Japanese, American and European companies in Europe was not just a matter of financial imperialism, however; it also carried long-term strategic connotations *and* implications for the industry's evolving geography as new and different patterns and processes of localisation began to emerge.

3.4 Concluding Comments

By the early 1990s, most of the truly major automotive component companies active in Europe - whether European or American-owned - operated on an international basis within the continent, in a fashion which mirrored the activities of leading vehicle assemblers. Production strategies were cast at this scale in order to attempt to meet a number of different and not always complementary needs: establishing and maintaining close technical and commercial links with car manufacturers wherever they were located, reaping the benefits of low-cost production locations, and satisfying different national political pressures. In a situation where market share, and especially leadership, conveyed improved returns on capital investment in order to finance further and increasingly-significant development programmes, it was vital to capture supply contracts not just in one or a few countries, but across Europe.

For the products in which these companies specialised, the same pressures also applied not just within Europe, but on a global scale. This was one factor behind American direct foreign investment in the sector in Europe, which in turn drew strength from these leading companies' overall financial size. Put simply, their absolute resources were far greater than those of practically all European companies, with the possible exception of Bosch. Their auto divisions were major producers *both* in Europe and in North America, whilst by contrast nearly all European companies were heavily dependent on Europe. Additionally, Japanese component companies - initially unchallenged in their home market (which gave them a head-start in global terms) - expanded into North America in the 1980s, and began to push into Europe during the 1990s. Whilst small by comparison with leading US firms - with the exception of Nippondenso - these companies were nonetheless massively strengthened by their association with one or other leading parent enterprise. On this basis, the position of many European companies looked increasingly uncomfortable.

Within this emergent global agenda, UK companies occupied an even more awkward role. Their market share and technological quality within chosen (and narrowly-defined) product ranges were high, as evidenced by contracts with the new Japanese assembly plants. On the other hand, their overall size on a European scale, still less by global standards, was relatively small. This raised crucial long-term questions about their ability to weather cyclical downturn, and to generate sufficient financial resources to sustain long-term development and business renewal programmes. Their position in the period to the late 1990s was potentially rendered even more problematic by the prospect of further Japanese investment in the automotive components sector in Europe and/or of further European and American investment in the UK. It was no coincidence, for instance, that Bosch and Valeo invested heavily in the UK in the late 1980s. Significantly, too (especially for the long-term) much of this was in the form of "hollow" manufacturing, with limited research and development capacity. These questions were important on a strategic basis, for although there were numerous niche product markets within which smaller companies could specialise, only the very biggest companies were likely to be able to develop and improve the high-technology content core products.

In this way, on-going processes of international competition within Europe influenced (and took shape from) the outcome of struggles within particular product ranges by European, American and Japanese corporations. This encounter, however, was heavily conditioned by differing regulatory conditions imposed by national states. Within Europe there were some evident contrasts. On the one hand there was the generally supportive attitude of the French government to Valeo's efforts at consolidation in the automotive component sector there, and at expansion to become something like a European "champion" (and similar attitudes towards Magneti Marelli and Fiat were found in Italy). On the other, the UK government's macro-economic measures in the late 1970s and early 1980s, combined with the legacy of earlier policies which offered limited incentive to consolidation and internationalisation, effectively humbled much of the UK-owned components sector by the mid-1980s. To compound matters, even the larger and more profitable concerns in the UK were limited from further expansion, particularly by acquisition - as the government's refusal to sanction GKN's attempted purchase of AE made only too clear. On a global scale, 'national' interest was generally protected to a greater extent: particularly so in Japan, but also in the USA. International competition, in other words, was heavily mediated by differing national state attitudes and agendas.

It was in this context that an emergent debate within the UK over locational changes in the component sector associated with Japanese assembly plant investment (in turn only one part of a broader process across Europe) had to be situated. Arguments that JIT production carried "new" spatial connotations and locational necessities were as falsely deterministic as earlier neo-classical industrial location theories. There were changes to the automobile production process in Europe taking place as a result of Japanese investment and the responses of existing manufacturers. These were only one part, however - both constitutive element in and result of - changing international strategies and forms of localisation of component production and supply. In the interlocking of different production systems and national state policies lay the emergent shape of future patterns of component production and supply, within Europe and beyond. "Europeanisation" in the components sector, then, was simultaneously part of a global process of competition between companies, of occasionally competitive national state policies, and changing forms of localisation of economic activity in space. The supply contracts of Japanese vehicle manufacturers in the UK in the early 1990s pointed to the creative capacity of such capital flows and state policies; the decimation of the West Midlands economy in the early 1980s to their potential destructive power.

Chapter 4
The Japanese, the European Market and the Automobile Industry in the United Kingdom

Ray Hudson
Department of Geography, University of Durham, United Kingdom

4.1 Introduction

Recent investment by Honda, Nissan and Toyota in car production in the UK has occurred against a background of a long period of absolute decline in the UK car industry and a shorter period of absolute and relative decline in the car industry there. A falling share of world output became accompanied by an absolute decline in production from over two million cars in 1970 to under one million in the mid 1980s, despite successive attempts to arrest it (see Wilks, 1984). In 1968, encouraged by Government, the remaining nationally-owned car companies merged to form the British Leyland Motor Company (see Central Policy Review Staff, 1975). When this failed to produce its intended effects, it was nationalised in 1975 to ensure the survival of a company of pivotal importance to the balance-of-payments position, national industrial production and employment (see Department of Industry, 1975). Nevertheless, the decline of BL continued, seemingly inexorably. One reason for the continuing problems of BL, and more generally of car production in the UK, was that from the 1950s the car plants became the location of a recurrent struggle between managers and workers (for example, see Beynon, 1973). Capital:labour conflict had important consequences for corporate competitiveness. Whilst BL's position continued to deteriorate, other companies increased their market share but in the context of a dramatic reduction in national production and rapidly growing import penetration.

After the 1979 general election, however, Mrs Thatcher's Conservative Government took a very different view to that broadly shared by Governments over the preceding thirty years as to how to tackle problems of economic decline, production and industrial relations (see Gamble, 1988). In the case of the car industry, the Government's assiduous pursuit of inward investment from Japan was seen as the route to a qualitative transformation of production strategies and industrial relations. Moreover, British consumers buying Japanese cars made by British workers rather than - say - German cars assembled by Turks was a solution in tune with certain strains of the populist and nationalist appeal of the Thatcherite project. It aligned the interests of British consumers, car workers and the UK Government with those of Japanese car companies against those of other European companies, countries and workers.

If Japanese inward investment in car production was seen as bringing manifold advantages to the UK Government, it also had perceived attractions to the Japanese companies. At one level, this was simply one element in an expanding internationalisation of production in response to growing labour shortages and rising labour costs within Japan (Economist Intelligence Unit, 1992a). Within western Europe, a combination of average labour productivity, trades union passivity, and low wages made the UK a particularly attractive location. Moreover, the UK Government was prepared to offer generous financial incentives and political support to attract Japanese investment. A third reason is that the UK provides a production base within the EC. This grew in significance with the approach of 1992 and the Single European Market - which will in principle remove the capacity of national Governments to regulate access to their markets for commodities produced within the Community. This created political controversy within and for the EC, as the UK Government's pugnacious defence of the interests of Japanese producers within its national territory engendered conflict with other producers and national Governments notably those of France and Italy which imposed import quotas or negotiated voluntary export restraints. Fourthly, by the second half of the 1980s the appreciation of the yen on international currency markets was of growing significance. This was especially so for lower priced models, with the European and USA markets increasingly served by production there, whereas higher priced models could continue to be produced in Japan and exported from there to these markets. The appreciating yen led to other problems, however. It increased pressures on Japanese producers in Europe to reduce component imports from Japan and thereby raise "local content" so that the rationales of the economics of production and the politics of international trade converged.

These various changes had a variety of effects on car production and trade in the UK in the 1980s. Declining national production but rapidly rising car sales financed by a massive credit boom and income taxation changes resulted in imports growing enormously whilst exports faltered (see Hudson and Williams, 1989). A position of approximate balance at the start of the 1980s became a trade deficit of £6.6 billion on motor vehicles by 1989, over 30 per cent of the total UK visible trade deficit in that year. This deficit subsequently narrowed as imports fell and exports began to increase. Indeed, by the last quarter of 1991, the motor vehicles balance of trade was again in surplus as domestic demand declined sharply and production recovered modestly to over 1.2 million cars in 1990. This was linked to an even more spectacular transformation of labour productivity. This rose by 66 per cent between 1980 and 1987, primarily because employment fell very sharply, by 45 per cent (some 185,000 jobs) in car assembly and component manufacture. The pace of job loss further accelerated in the early 1990s, with a reduction of 70,000 jobs in the period between April 1990 and March 1992, according to estimates from the Society of Motor Manufacturers and Traders. This restructuring was only marginally a direct consequence of the activities of Japanese producers within the UK and primarily a result of changes within existing plants in response to the perceived threat posed by the Japanese. In the 1990s, a further transformation is in prospect as a direct result of the build-up of Japanese production, with output restored to early 1970s levels at around two million cars per annum. But this will largely - if not wholly - be a car industry within the UK rather than a UK car industry, with a very different geography of production and with production approximately equally divided between Japanese, USA and European groups.

4.2 Competition Between Automobile Companies in the United Kingdom

By the 1990s there was a sharply emerging dichotomy in terms of labour productivity between the new actual and projected car plants of Honda, Nissan and Toyota and the existing plants of Ford, Peugeot, Rover and Vauxhall within

Table 4.1 Total wage costs[1] in selected countries for the automobile industry, 1990 and 1991 (DM per hour, current prices)

	1980	1990	1991
Sweden	28.6	43.72	45.41
Germany	26.9	41.87	44.47
USA	24.8	32.07	35.05
Belgium	28.1	31.83	
Italy	17.1	31.67	
Netherlands	23.3	31.20	
Japan	14.5	28.64	33.87
Spain	12.6	28.43	
France	19.7	26.01	26.43
UK	15.0	25.58	26.64

[1] includes social costs as well as wages paid directly to employees.
Source: German Automobile Industry Federation (cited in the Financial Times, 25 March 1991 and 20 February 1992)

Table 4.2 Labour productivity in car companies in the United Kingdom, January - June 1991

	Employees	Production	Cars per employees
Nissan	2,900	62,929	21.7
Vauxhall	10,943	142,547	13.0
Rover	36,000	229,086	6.4
Ford (exc. Jaguar)	44,000	196,005	4.5
Toyota (projected)	1,700	50,000	29.4

Note: Nissan and Vauxhall only assemble cars in the UK. Ford and Rover are involved in full manufacture and assembly operations. Assessing productivity in terms of car per employee is problematic (for example, production methods and the degree of subcontracting can vary greatly) but there is no readily agreed upon alternative measure.

Source: Society of Motor Manufacturers and Traders and Company estimates (cited in Independent on Sunday, 3 September 1991)

the UK. Despite - or perhaps because of - the UK having the lowest wage costs amongst the major automobile producers of western Europe (Table 4.1), non-Japanese plants in the UK were characterised by low labour productivity and quality of product (for example, see Table 4.2).

This was not, however, simply a question of productivity lagging behind the actual and projected levels in new Japanese plants. Within multinationals with plants in both the UK and continental Europe, there were also yawning gaps in productivity and quality. This is clearly illustrated by the case of Ford (revealed in parts of confidential internal reports selectively leaked and reported in the Financial Times). This remained the case although Ford had been acutely aware of this situation for some time, pursuing its "After Japan" strategy in an attempt to secure its position in world markets. In 1989 a confidential internal report revealed enormous differences in labour productivity and product quality between its UK and continental European plants (see Table 4.3). Dagenham in particular was revealed as the least efficient of Ford's European assembly plants. Despite strenuous efforts to remedy the situation serious problems persisted. A further confidential internal company report in 1991 referred to the continuing "yawning gap" in labour productivity between Ford's UK and continental European plants. Consequently, despite lower wage rates, Ford's costs per car were much higher in the UK, a situation that was expected to continue during 1991. Clearly, whilst Ford's problems were most severe at Dagenham, they were generalised across its UK operations and labour productivity within Ford's UK plants continued to lag behind that in its European plants. Moreover, growing integration of Ford's production system within Europe led to the effects of problems within its UK plants being rapidly transmitted to its other European

Table 4.3 Labour productivity and cost differences between Ford's European plants, 1989-91

Vehicle/ Location	Hours per Vehicle[1] 1988	1989	1990	1991-forecast	Costs[2] 1991-forecast
FIESTA					
Cologne		33.5	29.9	28.5	1,084
Valencia	33.0	35.0	33.3	34.1	912
Dagenham	57.0	59.4	52.2	40.0	1,344
ESCORT					
Saarlouis		29.5	33.9	30.5	1,070
Halewood	59.0	58.7	63.8	43.0	1,481
SIERRA					
Genk	40.0				
Dagenham	67.0				
TRANSIT					
Genk		48.7	45.7	45.1	1,445
Southampton	79.0	70.4	73.2	64.0	1,862

[1] based on salaried and hourly paid employees

[2] labour and overhead costs, $ USA

Source: Internal Ford Reports (cited in Financial Times, 26 June 1990 and 23 December 1991)

factories. A seven week strike at Halewood during 1990 produced a rapid "domino effect" that revealed the extent to which Ford's European production was vulnerable to industrial action in one UK factory. The "unreliability" of supplies from its UK plants because of industrial action was soon after cited as a major reason for Ford switching a planned £225 million investment in phase two of its Bridgend engine plant to Cologne. Clearly, while strikes and other forms of industrial action in the UK were much reduced in comparison with previous decades, adversarial industrial relations within its UK plants in the 1980s, compared to those of its continental European ones, were leading Ford to rethink the geography of its European production strategy and the place of its UK plants within this.

Ford most sharply exemplified the problems of car production in the UK by the late 1980s/early 1990s. Nevertheless, if other established car producers in the UK were in a relatively favourable position *vis a vis* Ford at this time, they too were seriously threatened by growing productive capacity in the Japanese plants (Table 4.4). Womack et al. (1990) suggest that European car companies on average take twice as many hours as Japanese ones to assemble a car, double the engineering effort to develop a new one. The arrival, actual and impending, of Japanese plants simply intensified pressures on existing ones in the UK, both

Table 4.4 Japanese automobile makers investments in the United Kingdom: Actual and projected impacts

Company	Location	Functions	Investment £ million	Planned capacity	Actual production	European component purchases £ million	Jobs
Nissan	Sunderland	Assembly and engine plant	850	300,000 (1993)	124,000 (1991)	850 (1993)	4600 (1993)
Nissan Yamato (80 per cent Nissan)	Sunderland	Heavy pressings	50				
Nissan European Technology Centre	Cranfield	R and D	31				250 (plus 100 at Sunderland
Nissan Motor GB	London and Lutterworth	Sales,marketing & distribution	40				375
Toyota	Burnaston	Assembly	700	200,000 (1997)	Began late 1992	700 (1997)	3,000 (1997)
	Deeside	Engine plant	140	200,000 (1997)			30
Honda	Swindon	Assembly	300	100,000 (1995)	Began late 1992		1,400
		Engine plant	62	120,000 (1995)	80,000 (1991)		600

Source: compiled from press reports

those of Rover and those of multinationals with more productive facilities in continental Europe, to which they could switch production.

Nissan was the first Japanese company to assemble and then produce cars in the UK and the significance of its decision should not be underestimated (see Beynon, 1986; Crowther and Garrahan, 1988; Garrahan, 1986; Garrahan and Stewart, 1992; Sadler, 1992a). From its modest beginnings in the small scale assembly of KD kits in 1986, production in 1993 reached 246,000 cars, of which 75 per cent were exported. As early as 1989 Nissan's President, Mr. Yutaka Kume predicted "400,000 cars a year in the UK by the late 1990s". Furthermore, as well as expanding productive capacity, Nissan has significantly broadened its operations in the UK in two ways. Firstly, via establishing its European Technology Centre at Cranfield : its stated aim, is that Nissan "should be able to design a vehicle from scratch in Europe" by the mid-1990s (Sadler, 1992a). Secondly, via establishing its wholly-owned UK distribution network. This was intended in the short-term to recover its UK market share to 10 per cent in 1992, following a well-publicised dispute with its former distributor, and to increase it further beyond that date. It also signalled a change in Nissan's marketing strategy, from one based on price competition to one that emphasises performance, quality, reliability and technology. With the launch of its own dealer network in 1992, Nissan became the only foreign car company to design, manufacture, market and sell cars in the UK (although its European HQ is located in Amsterdam).

In this way, Nissan is seeking to resolve the "global/local paradox facing all car manufacturers", as Robert Stempel, Chairman of General Motors, put it at the 1991 Financial Times World Motor Industry Conference. Stempel stressed that there is not a single "world car" despite all the "daring talk" of the last decade, but rather a need to grapple with the global/local paradox. The essence of this paradox is that companies must be global in outlook, utilising ideas, materials and designs from all over the world to meet customers desires and sourcing components globally while at the same time thinking locally, as tastes and conditions differ too widely for a single car to satisfy everyone. Resolving this paradox is precisely what Nissan has done at its Sunderland plant but this in turn must be seen as the European component of a global strategy. As the company pointed out in its 1989 Annual Report:

> Today, Nissan is taking globalisation much further through a long-term localisation programme - which involves the creation of bases in key markets to coordinate all regional research and development, procurement, production, marketing and finance in line with local conditions....Nissan has long been Japan's most globally active company. Localisation is just a logical extension of this.

Honda and Toyota will also undoubtedly seek to resolve this global/local paradox in their own ways at and around their UK plants.

Unlike Nissan in the first two phases of its development at Sunderland, Toyota received no financial aid from the UK central Government. This was because of an EC regulation, introduced in response to protectionist pressures, which required state aid in excess of £8 million to automobile companies to receive prior EC approval (see Sadler, this volume). Seeking aid would have delayed the project at a time when Toyota was concerned long-term with market share within the immanent Single European Market, and short-term with its competitive position *vis a vis* Nissan, so it chose to proceed quickly without it. Production began in 1992 and then increased rapidly as intended (Table 4.4). Following Nissan, Toyota too then announced plans to establish its own distribution network.

Following its long-running collaboration with BL/Rover (see below), Honda announced plans in 1989 (shortly after Toyota announced its plans) to invest in assembly and engine plants at Swindon on a site bought in the mid-1980s. Honda made clear its commitment to produce 100,000 cars per annum there by 1994. However, further capacity increases there cannot be discounted since a minimum annual output of 200,000 cars is emerging as the norm for new plants. In these circumstances, and given the continuing importance of scale economies, Honda will not wish to be disadvantaged by a sub-optimal level of production.

Leaving aside possible future increases by all three Japanese companies, on the basis of known investments and productive capacity they will be producing around 600,000 cars per annum by the late 1990s, in the ratio 3:2:1 as between Nissan, Toyota and Honda. If annual car production in the UK recovers to the level of the early 1970s as a result, it will be split approximately equally between companies based in Japan, the USA and Europe. The competitive threat posed by the Japanese to other producers within the UK is therefore severe in both domestic and export markets. For whilst much (80 per cent in the case of Nissan and Toyota) of the output of Japanese plants in the UK is intended for export to the rest of the EC, these same companies will continue to import finished cars into the UK market, from the USA as well as Japan. Moreover, there is clear evidence that Japanese producers will sacrifice short-term profits for long-term market share. As the 1991 Euromonitor Financial Yearbook noted, every car produced by Nissan at its Sunderland plant had been sold at a loss. Whilst this may reflect internal transfer pricing decisions, it also reflects decisions on market pricing.

Fierce competition *between* the Japanese producers may well lead Honda and Toyota to follow similar strategies. Furthermore, new Japanese capacity will be coming on-stream against the background of collapse in the national market for new cars, as the UK economy slid into a deep and persistent recession in the late 1980s and early 1990s from which it shows little sign of emerging quickly. This has had a sharp impact on those producing in the UK for the UK market, as imports continued to account for over 50 per cent of sales in the early 1990s.

Whilst the non-Japanese manufacturers of volume cars suffered relatively much less than those of luxury marques, growing competitive pressures triggered further cuts in employment in search of greater competitiveness via mimicking Japanese "lean production" methods. There had been significant changes in the market shares over the 1980s as the national market expanded to

Table 4.5 Market share in the United Kingdom, 1981 and 1990

	1981	1990
	Percentage	
Ford	30.9	25.3
Rover (formerly BL)	19.2	14.0
Vauxhall (GM)	8.6	16.1
Peugeot (inc. Citroen and Talbot)	7.7	9.2
Nissan	5.9	5.3
VW (inc. Audi)	5.4	6.4
Renault	4.9	3.4
Fiat (inc. Lancia and Alfa Romeo)	4.7	3.1
Volvo	3.0	3.3
Others	9.7	13.9

Source: Society of Motor Manufacturers and Traders

record levels; the positions of Ford and Rover deteriorated sharply whilst that of Vauxhall improved greatly (Table 4.5). This competitive struggle intensified as the national market subsequently shrank sharply. This is not just a competitive struggle between the Japanese and non-Japanese producers or between the non-Japanese producers as they seek to survive, however, but also one between the Japanese producers themselves. Moreover, for them, this is not so much a struggle for domination of production in the UK and the UK market as for shares of the massive EC market, the largest in the world. This has been apparent from an early stage.

4.3 Cooperation Between Automobile Companies in the United Kingdom

As well as competition, there has also been growing cooperation, in the form of strategic alliances, between some car producers in the UK. In one sense, this is no more than a reflection within the UK of a growing global tendency towards such alliances, involving even the largest automotive producers, to help cover the escalating costs of R and D, product development and so on. Indeed, such alliances can be seen as a *specific* form of globalisation that burgeoned in the 1980s (OECD, 1990). Clearly, whatever the validity of claims about flexible specialisation, economies of scope and burgeoning numbers of small firms as growing tendencies in other sectors, mass production, economies of scale and yet-bigger firms remain vitally important within the automobiles sector as old tendencies towards centralisation and concentration of capital take new forms.

Within the UK the most significant strategic alliance has been that between Honda and Rover, a relationship that began in 1979. For Rover, then part of British Leyland, it offered access to Japanese production technology; for Honda, it offered access to UK and EC markets. The continuing troubles experienced by the Rover Group resulted in privatisation via its sale to British Aerospace in 1988 for the remarkably low price of £150 million, following an injection of £547 million of public money into Rover by central Government (National Audit Office, 1989). Privatisation of Rover opened the way for a significant extension and deepening of its collaborative relationship with Honda. The Japanese company took a 20 per cent equity stake in Rover and the latter began to produce Honda models at Longbridge. Later in 1991 further intensification of the links between the two companies extended their strategic relationship to the end of the 1990s.

Enhanced collaboration with Honda offered Rover the possibility of changing from one of the least to one of the most productive car makers in Europe. Such a transformation, if successful, would carry a heavy price, however: Rover would become increasingly dependent upon Honda for know-how and production technology. By the mid-1990s all mainstream Rover car model programmes were planned to be based on collaboration with Honda. The 1991 Euromonitor Financial Yearbook suggested that by the end of the 1990s, Rover's interests in the car industry might be limited to maintaining its name on Honda-manufactured products and holding a minority stake in Honda's European operations. Whilst there may have been continuing uncertainty surrounding the ownership of Rover's parent company, British Aerospace, and of Rover itself (since it became possible for British Aerospace to sell its stake in Rover to a non-UK company from August 1993 without financial penalty), Rover's development as a car producer seemed inextricably tied to Honda.

What then was the attraction for Honda? Above all else, it was intended to give it a European presence to rival Nissan and Toyota. The deal immediately provided more productive capacity to help cope with the 1991 trade agreement, limiting further Japanese penetration of the EC market to cars produced within the EC. It also gave it time to develop its European dealer network whilst building up production at Swindon. But as well as short-term tactical advantage, there were long-term strategic considerations. If Rover achieved the required levels of efficiency, productivity and quality both it and Honda would reap great competitive advantage as a result of shared economies of scale generated by collaborative product development and a joint supplier base, established under Honda's leadership to world class cost and quality standards. If Rover failed to make this qualitative transformation, however, Honda had an alternative in the form of its own new Swindon assembly plant. All of its European production will be concentrated there. A 1989 proposal for half of the output of this plant to be Rover's had been dropped. Rover was to produce its share of the joint model range agreed in 1989 at Cowley and cease to assemble Hondas (up to 40,000 per year) at Longbridge. If Rover failed to achieve the desired and required transformation in the 1990s, Honda's Swindon plant would remove the risk of experiencing productivity and quality problems at Rover and enable it to compete with Nissan and Toyota on their greenfield sites.

All of this was suddenly altered in January 1994, when it was unexpectedly announced that BMW was to purchase British Aerospace's 80 per cent stake in Rover. This raised considerable question marks over Rover's future and fears that it might now become little more than an assembler of automobiles designed in Germany. It also led to uncertainty about Honda's European strategy, since it was clear that it had little notice of BMW's interest and was strongly opposed to it taking a majority shareholding in the company. In the months following the takeover, BMW sought to allay fears about Rover's future as an independent company and to avoid a dramatic rupturing of relations with Honda and ensure that the Honda-Rover relationship continued, though in a diminished form. It also became clear that, for Honda too, there were good commercial reasons for ensuring a continuing relationship with Rover for the forseeable future. It is not yet clear however as to the precise pattern of new relationships that will emerge between the three corners of the BMW, Honda and Rover triangle.

4.4 Cooperation Between Component Suppliers and Automobile Producing Companies in the United Kingdom

Historically, relationships in the UK between car companies and their component suppliers have been antagonistic. The component sector has been characterised by a plethora of small firms, competing vigorously on the basis of price for contracts with much larger car manufacturing companies. This was a very different institutional structure to those characteristically found both in the USA and in Japan. Prior to the arrival of Japanese plants in the UK, however, but influenced by competition from Japanese producers in the world market, car manufacturing companies in the UK were seeking to restructure relations with their suppliers along the lines of the system found in Japan. Typically, in Japan, a few large "first tier" suppliers account for more than 60 per cent of the value-added to a vehicle in the production process, with thousands of small "second" and lower tier companies supplying them. From the early 1980s, BL and Ford were pressuring component companies in the UK to improve performance and efficiency, offering fewer but larger contracts to suppliers able to meet strict quality and financial targets (see Amin and Smith, 1990; 1991).

Thus restructuring of the component sector within the UK onto a more hierarchical basis, with greater differentiation between first and lower tier suppliers, had begun prior to the arrival there of Japanese car producers. Equally, this process has undeniably been accelerated by the demands of the new Japanese assembly plants for total quality assurance from their component suppliers.

From the outset, both Nissan, Toyota and, more recently, Honda committed themselves to "local" - European - component sourcing. By 1986 Nissan had 67 European suppliers; by 1990 it had 177, of which 120 were UK companies accounting for 74 per cent of European purchases by value. Only five of these Nissan suppliers are Japanese owned. Eight are European-Japanese joint ventures and twenty five have technology agreements with Japanese suppliers. But, perhaps significantly, until recently Nissan coyly refused to disclose the value of components it continues to purchase from Japan, which include and will continue to include gearboxes and engine components such as blocks, crankshafts and con rods. But in September 1992 Nissan revealed that in 1993 it expected to purchase £850 million of components within Europe, an increase from £220 million in 1990, and to import components to the value of £240 million from Japan. Toyota apparently has an even stronger commitment to European sourcing, insisting that its suppliers will include only a handful of Japanese companies. The vast majority of its suppliers will be based in the UK and continental Europe. Honda, too, recently announced the list of 136 component companies "appointed" to supply its Swindon plant. Of these 89 are based in the UK; 44 in continental Europe (predominantly France - 19 - and Germany - 13) and two in the USA.

If anything, then, Honda and Toyota will be even more vulnerable to any shortcomings in the European components companies that they choose as their suppliers. The processes through which these companies have come to select their component suppliers are, therefore, of particular significance. They are best illustrated by the case of Toyota. Early in 1989 staff in Toyota's European administration office in Brussels compiled a list of over 2,000 potential suppliers. Closer scrutiny reduced this to 400 companies, serious contenders for contracts. Already this partial screening had reduced Toyota's potential suppliers to fewer than half the average number of actual direct suppliers to most European vehicle producers. Having reduced the list to 400, the selection process became much more painstaking and demanding of potential suppliers. Over a period of ten months, multi-disciplinary Toyota teams assessed their capabilities in relation to four key criteria : management attitude and capability; manufacturing and production facilities and their level of technological investment; quality control philosophy and systems; and R and D capability. This further reduced the list of potential suppliers to 250. It was only at this stage that companies were asked to submit costs, "to give us a feel for price competition". For Toyota, like Nissan, price is not *initially* the main criteria for selecting component suppliers (NEDO,1991). All were deemed acceptable in terms of their potential to meet quality and price standards and the first prototype parts were ordered in October 1990. This does not mean, however, that price has ceased to be an significant dimension of competition; quite the contrary, for the Japanese companies subsequently expect to benefit from price reductions as a result of cost savings by their appointed suppliers (see below).

Toyota remains cautiously optimistic about the extent to which the qualitative transformation that it demands from its suppliers actually is taking place. Cautious optimism is probably the most appropriate reaction, for NEDO (1991) reported that Nissan rejected at least 1 in 1000 components at Sunderland compared to 1.2 in 10,000 in Japan :"....the UK components industry still has a long way to go....the position in the UK, while good by our standards, is a generation behind Japan". Indeed, it also lags behind continental Europe : if, on average, component quality in Japan was rated at 100, that in continental Europe rated 80 while that in the UK would only score 65-70.

Toyota, like Nissan before it, strongly emphasises the need for partnerships with its suppliers. Consequently, it too has established "technology help teams", which, if necessary, go into suppliers' factories to explain aspects of production systems and help solve problems relating to them. A fuller picture of what giving assistance means in practice for relationships between car producers and component suppliers, and in particular what it requires of the latter, emerges from the NEDO (1991).

Several key features of the relationships between Nissan and its suppliers echo the type of relationships it expects in Japan. First, the supplier has to regard itself as part of a team and undertake long-term commitments. Secondly, it has to guarantee frequent and precisely-timed parts deliveries. Thirdly, close collaboration is required on component design and development, irrespective of company size. Fourthly, suppliers are placed under continuous pressure to improve. Nissan focuses on suppliers' production costs and works with them to achieve efficiency savings. Finally, there is also persistent pressure from Nissan for continuous price reductions, often based on detailed breakdowns of price and cost. This has been, and continues to be, a source of tension between Nissan and its European component suppliers, especially those in the UK, which have difficulty in understanding why Nissan thinks it equitable that suppliers should pass on most, if not all, of cost savings achieved through enhanced efficiency.

While Japanese companies are seeking stable long-term relationships with their suppliers, these are not, however, unconditionally and permanently guaranteed contracts. Their continuation depends upon component companies meeting exacting delivery, price and quality criteria. Japanese car producers in this way seek to promote competition between component companies so as to maintain those inter-company relationships that best serve their own interests and competitive position. Collaboration is therefore based upon profoundly asymmetrical power relations between the assembly companies and their component suppliers.

It is, though, also important to note that there are dangers to the car assemblers as a result of growing reliance upon just-in-time component supply strategies. "Japanese-style management" and "just-in-time production" involve much more than just stock control. The smooth reproduction of cooperative partnerships between assemblers and component suppliers depends upon tranquil relations between workers and managers within both assembly and component plants. This is above all a matter of "soft" cultural change. There is evidence that this is being achieved more successfully by the new Japanese companies, able carefully to select their workforces in localities with no history of car production, than by those established for much longer in Europe.

4.5 Competition and Cooperation between Component Companies in the United Kingdom and European Community

While the automobile component sector in the UK has been characterised by a plethora of small firms, half a dozen companies - Lucas, GKN, Pilkington, T and N, BBA and BTR - have annual turnover of at least £500 million; 7 per cent of component companies account for 60 per cent of employment. This concentration of employment and activity grew over the 1980s. The dichotomisation between the six biggest UK component companies and the remainder is symptomatic of a profound process of restructuring that began to redraw the corporate anatomy of the components sector. For not only was there a growing proportion of activity within the six largest companies but they were also becoming increasingly dominant and specialised in particular product markets, whilst changing the importance of automobile components within their overall mix of activities. Unlike some of their major continental European rivals but like some of those based in the USA, the major UK component producers are parts of diversified conglomerates, switching between production in the automotive and non-automotive sectors as an element in corporate strategies (see Sadler and Amin, this volume Table 3.1; Sadler, 1991a). In these sorts of ways, then, elements of the component supply chain were selectively re-defined via processes of product market segmentation and corporate product specialisation, as these companies emerged as prime UK "first tier" suppliers.

There has also been an increasing internationalisation of production as well as sales within Europe by UK component producers. This spatial shift was given great impetus by the deep recession in the manufacturing sector in the UK in the early 1980s. Faced with a severe contraction in car production within

Table 4.6 Foreign investment in the United Kingdom automobile components industry : some typical examples

Date	Company	Country	Product	Type of investment
1986	Ikeda	Japan	Car seats	Joint venture with Hoover
	Nihon Radiator (now Calsonic)	Japan	Exhaust systems	Joint venture with TI
1987	Magneti Marelli	Italy	Lighting, alternators electrical operations	Takeover of Lucas operations
1988	Arvin Industries	USA	Exhaust systems	Takeover of Cheswick and Bainbridge business from TI
	Yuasa	Japan	Batteries	Joint venture with Lucas Automotive
	Nippen-Seiki	Japan	Dashboard instruments	Greenfield
	Eisenwerk Brühl	Germany	Castings	Takeover of Austin-Rover foundries
	Devalit Plastic	Germany	Trims	Greenfield
	Plastic Omnium	France	Fuel tanks and bumpers	Greenfield
	Minebea	Japan	Bearings	Takeover from APV
1989	Valeo	France	Heaters	Takeover of Delanair
	Monroe Auto Equip.	USA	Shock Absorbers	Takeover of Armstrong Equip.Division
	Calsonic	Japan	Radiator	Takeover of Llanelli Radiators
	Nippondenso	Japan	Radiator	Takeover of IMI Radiators
1990	Magnetti Marelli/ Nippondenso	Italy/ Japan	Air conditioning	Greenfield joint ventures
1991	Bosch	Germany	Alternators	Greenfield

Source : Financial Times, various dates

the UK, UK car component companies both increased direct sales into, and established manufacturing capacity in, EC countries. The large and more buoyant West German market made it a particularly attractive target. At the same time, however, major European component companies were increasingly locating productive capacity in the UK, especially as it became clear that Japanese automobile companies intended to establish a major manufacturing presence there. This raised the prospect of a recovery in demand for components as Nissan, Toyota and Honda increased output at a time when the indigenous UK component industry was still reeling from the effects of a long period of retrenchment and decline. This inward investment took a variety of forms (Table 4.6). European (and other) component companies sought to take-over UK component manufacturers, for this provided ready entry to a country with low production costs and rapid investment in new car plants. Where acquisitions were either unavailable or unattractive, multinational components companies such as Bosch and Valeo established new greenfield factories. This growing penetration in varied forms by European component companies was one element in their overall global strategic response to change. Such plants were increasingly intended to serve the entire EC market in particular products and product ranges, as the importance of economies of scale continued to be a decisive influence on corporate strategies.

This emergent restructuring of the European components sector - the context in which the restructuring of the UK components sector and components production in the UK has occurred - must also be seen in relation to a potential threat from inward investment by Japanese component companies. There has already been some investment by such companies, via takeovers, in the form of joint ventures and licensing agreements and, less frequently in new greenfield factories (see Table 4.6). However, the most popular approach is licensing and joint ventures, with 30 of the 50 Japanese component companies operating within the EC in 1990 following this approach. To the Japanese companies, these appear as less controversial than solely owned Japanese plants on greenfield sites; even new greenfield plants may be joint ventures. This is especially so given the ongoing controversy about "local content" of Japanese cars produced within the UK/EC. To the participating European companies, joint ventures offer access to Japanese technology. Even in circumstances in which national Governments and "national" car companies are vehemently opposed to the Japanese presence, the lure of Japanese technology can prove irresistible.

To some, the continuing gap in performance and quality between the European, especially the UK, and Japanese component companies suggests an immanent danger of Japanese component companies flooding into Europe in a way reminiscent of the earlier invasion of North America. Despite a narrowing in the productivity gap between German and UK components companies in the 1980s, both continue to lag behind the productivity levels of Japanese companies (Boston Consulting Group and PRS Consultancy International, 1991). Moreover, Japanese component companies are not immune to pressures to globalize their activities in order to remain competitive and this could well lead them to seek to invest within the EC. Others, however, take a more sanguine view of the prospects for the European components industry. There are significant institutional differences in relations between assembly and component companies in Europe as compared to the USA. Moreover, the European components sector was already moving towards Japanese standards prior to the arrival of Japanese car producers in the UK. Add in the continued salience of the "local content" issue and the threat of an inflow of Japanese component investment similar to that in the USA can, some argue, be "largely discounted" (Economist Intelligence Unit, 1992b).

While there may be uncertainties as to the future, it is clear that the European, including the UK, component sector is currently increasingly being restructured around a few, big multinational companies, often multi-product conglomerates, which constitute the emergent "first tier" suppliers. It is they who are entering stable, long-term relations with car producers within Europe, though those whose horizons remain simply European are "already behind in the race" (Economist Intelligence Unit, 1992c). Below the "first tier" companies, however, supplying them, is a much larger number of smaller "second tier" suppliers. But these "second (and lower) tier" suppliers are discovering that their relationships with the "first tier" companies are no longer simply regulated by price competition in the market place. The "first tier" companies are seeking the same sorts of relations with their suppliers as the assembly companies have negotiated with - or perhaps imposed upon- them. This is not necessarily an easy task. Bosch, for example, encountered problems in finding suitable UK suppliers for its alternator plant at Cardiff. The inference is clear, however. The growing demands of "first tier" of suppliers will lead to a shake-out amongst their "second" tier suppliers; and so on down the supplier chain. This restructuring within Europe nevertheless still has some way to go. Furthermore, especially in relation to components for which the production process remains labour intensive, further restructuring may well lead to dramatic changes in the geography of production of major component companies not only within but beyond Europe.

4.6 Capital: Labour Relations

A central element of the Thatcherite project in the United Kingdom in the 1980s was to reduce the influence of trades unions (Fairbrother and Waddington, 1990). This was not a question of abolishing trades unions but rather radically to alter their character and practices. Great emphasis has been placed upon the significance of new industrial relations procedures within the car plants of Honda, Nissan and Toyota. There has even been talk of a process of 'Japanisation', which other companies are being forced to mimic in their UK and EC plants. This is a one-dimensional oversimplification of more subtle and multidimensional processes of change, however. Japanese companies have undeniably had a major impact but this has involved much more than a mechanistic transfer of attitudes and practices from Japan and their slavish

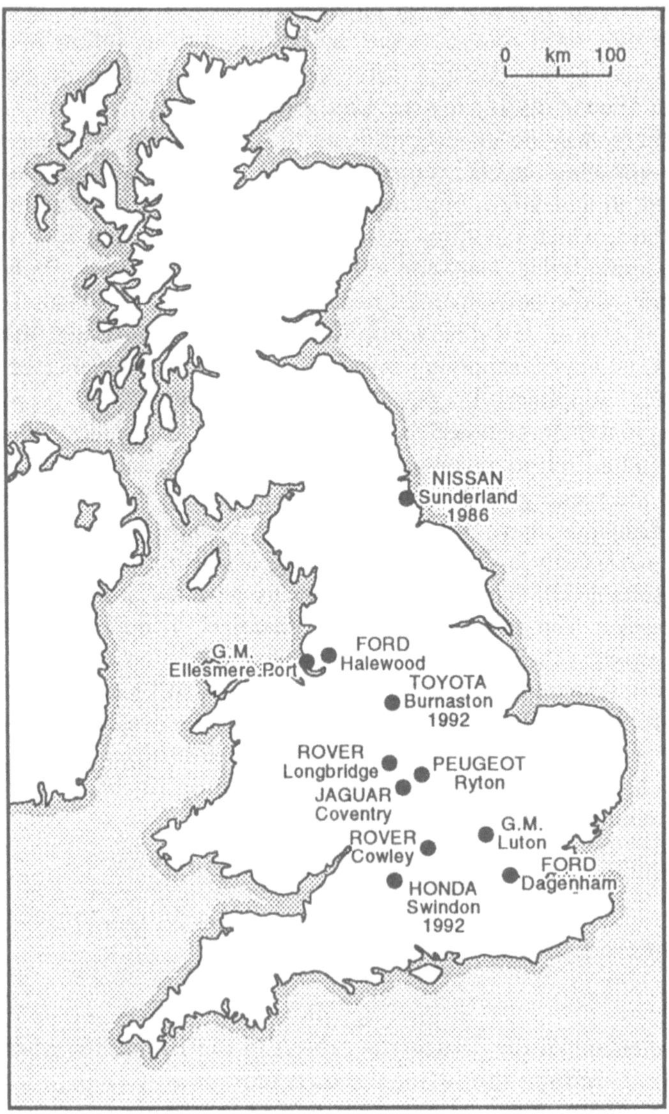

Fig. 4.1 Automobile assembly plants in the U.K.

imitation within Europe. A central element in the process of change has been the very careful selection of locations within the UK (Fig. 4.1), especially in terms of their labour market conditions and more general receptivity to Japanese investment there, coupled with a Japanese willingness to adopt, to a degree, to local conditions. This is clearly exemplified by Nissan's position with regard to a non-union or one-union plant at Sunderland. At an early stage Nissan made it clear that it would not insist on a no-union deal though it would prefer the number of unions to be as small as possible. Peter Wickens later (1987) stressed the significance of labour market conditions and the character of trades unionism in the north east of England in influencing Nissan's choice of Sunderland:

> none of the unions could be described as being on the left of union politics. The experience [of industrial decline] of the last few years...has resulted in the development of an attitude among trades union officials, that they must do everything possible to assist in attracting industry and commerce.

In coming to this view Nissan's Personnel Director had doubtless been influenced by voices such as those of Tom Burlison, chairman of the Northern Region of TUC (cited in Sadler, 1992a):

> When Nissan first came to the north east three years ago we said we would bend over backwards to help and our attitude remains the same. We want Nissan and will do whatever we can to bring about the kind of union agreement they seek.

Under these circumstances, Nissan could agree to a one-union deal with little difficulty, safe in the knowledge that, in the ensuing "beauty contest" between unions themselves desperate to recruit new members, it would secure the kind of one-union deal that suited its corporate purposes. Thus Nissan was able, in its own words, to "appoint" the Amalgamated Engineering Union (AEU) on terms that it could dictate.

Perhaps because of this rather restricted definition of permissible trades union action, perhaps because of Nissan's very careful screening of its workforce to ensure commitment to the company, the Sunderland plant remains relatively non-unionised; by 1991 the level of unionisation was 35 per cent, having stabilised at this level some time ago. Nissan, however, publicly stressed that it had an "excellent relationship" with the union.

Whilst Honda has, so far, decided against union recognition at Swindon, Toyota has followed a very similar approach to Nissan. Its apparently relaxed stance was projected from a position of considerable strength. At least five trades unions [AEU; Transport and General Workers Union (TGWU); Manufacturing Science Finance (MSF); Electrical Electronic Telecommunication Plumbing Union (EEPTU); and General Municipal Boilermakers (GMB)] were anxious to obtain sole negotiating rights, while Toyota had already commissioned market research by PA Inbucon into the unions on offer. In the end, Toyota came to a single union agreement not unlike that agreed by Nissan with the AEU. Clearly, then, management in the new Japanese car plants is not necessarily anti-trades union. What is at issue is the form of trades unionism and whether a one-union or non-union deal is seen as the most appropriate mechanism to secure the industrial relations climate and production regime desired within the factory. No-union or one-union status is not an end in itself, but rather a means to an end.

The case of Toyota exemplifies the care that Japanese companies exercise in selecting not only component suppliers but also trades unions to represent their workforces. They are equally assiduous in the selection of those workforces. They can exercise great selectivity because they have chosen to invest in locations with no previous experience of car production and high unemployment. Thus the car plants are swamped with applicants, especially as they are perceived to offer secure and often relatively well-paid jobs in relation to the areas around them. The recruitment process involves great care in choosing production workers but, even more so, supervisors and team leaders who play a pivotal role in regulating the labour process. The companies have been very clear from the outset as to the type of workers they were seeking. Nissan, for example, only recruits about 35 per cent of its workforce from the ranks of the unemployed. Locating in Sunderland minimises the chances of people with experience of working in car plants and Nissan clearly prefers to recruit those in work who exhibit appropriate attitudes and commitment to their new employer. Whether employed or unemployed, Nissan only recruits people aged 35 or less. In 1987, for example, the average age of its workforce was 27 years. This age selectivity is closely related to the physical demands of work on the shopfloor. As in other respects, Honda's and Toyota's recruitment strategies echo those of Nissan.

This careful, thorough and painstaking, recruitment strategy provides the basis for flexible, high productivity and high quality production within the car plants. A rigid "Japanese" managerial style is not necessarily imposed there, however; for example, compared to Honda in Swindon, Nissan in Sunderland has a managerial approach more reminiscent of Germany than Japan. The emphasis is upon blending elements of Japanese management styles with "best practice" from other systems and parts of the world, adjusted to give maximum effectiveness in the specific local circumstances of each plant.

Nissan has been extremely pleased with the performance of its Sunderland plant. Senior Japanese managers within Nissan have been fulsome in their praise. There is no doubt that this reflected the efforts of the workforce rather than heavy fixed capital investment in sophisticated production technologies and high levels of automation. Nissan is responding to increasing labour shortages and costs in Japan by investing in technically very sophisticated and automated new production methods in its factories there. They are incorporating a degree of flexibility in product mix; a form of flexible mass production and dynamic flexibility (see Coriat, 1991). In contrast, the Nissan plant at Sunderland does not represent a revolution in automation or robotics but it has intensified the pace of work under the rubric of 'flexibility' and set standards in working practices which pose a formidable competitive threat to other car plants in the UK and EC. Moreover, Nissan's workforce has been left in no doubt that it must maintain these high standards of quality and levels of labour productivity. The company is making no unconditional promises of secure employment for life at Sunderland, a point that was powerfully reinforced towards the end of 1993 when it became clear that Nissan was seeking to reduce its workforce at Sunderland in response to falling demand for new cars in its EC export markets. Whilst Nissan represents its working practices at Sunderland as empowering workers via its integration of quality, flexibility and teamwork, Garrahan and Stewart (1992, 111) describe them as a new regime of subordination, based around control, exploitation and surveillance, that disempowers workers and which in many ways closely resembles the practices of Fordism (op. cit., 125).

One consequence of the revolutionary - by UK standards - gains in labour productivity achieved at Sunderland by Nissan has been mounting managerial pressure for parallel changes at other car plants in the UK and EC. In the mid-1980s Ford claimed that Nissan would reduce production costs by between £330 and £530 per car because of its new working practices (Sadler, 1992a). The prospect of further productivity increases promised by Honda and Toyota at their plants further magnifies such pressures. Not surprisingly, given Honda's growing involvement in its operations, Rover has been prominent amongst those seeking radical changes in working practices. In 1990, for example, it announced its intention to introduce 24 hour a day, 7 day a week working amongst production workers. It was the first UK vehicle manufacturer publicly to declare that this was its objective. Eighteen months later, in a document entitled "Rover Tomorrow - The New Deal", it revealed proposals for major changes in working practices, specifically in response to the challenge posed by the new Japanese factories in the UK. In March 1992 Rover workers voted by a narrow majority to accept the proposed changes, though the large numbers voting to reject them raise doubts as to how easily change will be accomplished. There is, however, evidence from other Ford and Vauxhall plants that, given the choice of redefined jobs or the prospect of no job at all, workers and trades unions will, albeit reluctantly, concede changes. Moreover, continuing labour productivity differentials between Japanese and non-Japanese plants generate pressures for yet more profound changes in industrial relations and working practices, changes that may sharply alter the geography of car production within Europe. Similar pressures are transmitted to component supply companies. Clearly the influence exerted by the new Japanese plants, directly and indirectly, is extending throughout the entire car production *filière* in the UK, downstream from the assembly plants to component suppliers and beyond, and upstream from the assembly plants to parts and after-sales service activities.

4.7 State Regulation of Japanese Competition: European Community Trade Policies, National Interests and Corporate Interests

Japanese car production in the UK has grown in the context of the Japanese already taking about 10 per cent of the EC and west European markets via direct imports. In non-EC countries without quota protection and/or domestic car producers import penetration rose to between 25 per cent (Sweden) and 42 per cent (Finland) by the end of the 1980s. With the prospect of the Single European Market removing national restrictions on intra-EC international trade, such figures conjured up dire images and provoked extreme fears in EC countries with substantial domestic car production capacity. This was a fortiori so in countries currently heavily protected from Japanese competition via strict control of imports, notably France and Italy (see Sadler, this volume). For their national Governments and their major national car companies - Fiat, Peugeot and Renault - the threat of significant Japanese market penetration via exports from the UK is very worrying. Consequently, there is great political pressure on the EC Commission to regulate the overall Japanese share of the market, both via imports from outside the EC and from exports from the "transplants" in the UK to other EC countries. It also has resulted in strong demands to define and regulate "local content" and maintain this at high levels. Such issues continue to pose considerable policy dilemmas for the EC, torn between competing corporate and national interests (see Dicken, 1992a). The purpose of this section is not fully to review this situation and the EC's policy response to it (for this see Sadler, 1991b and this volume) but selectively to focus on some aspects of it within the UK.

Whilst Nissan was the harbinger of the Japanese assault, the announcement by Toyota, the world's third largest and Japan's largest automobile producer, of its plans for UK production signalled an escalation in the challenge to existing producers in Europe. This merely emphasised the vacuum in policy making at EC level for dealing with the looming wave of Japanese investment in the car industry. The Community was caught between strongly conflicting pressures from opponents and supporters of Japanese direct investment. Much of this conflict came to centre around the issue of "local content", the extent to which cars made by Japanese companies in the EC could be considered EC cars, and the imposition of some form of local content regime on the new greenfield Japanese car plants within the EC. Since this to all intents and purposes meant greenfield plants in the UK, it created complex tensions between the UK and other EC national governments and the UK government and other car producers within the UK.

But in fact "local content" is only symptomatic of a broader political conflict and policy debate surrounding the desirability or otherwise of the new Japanese car plants. This has aligned the UK Government, defending Japanese production within its national territory, against a variety of opponents within the EC. The UK Government insists that Japanese cars produced within the EC (in effect, the UK) should be excluded from any future restrictions on imports into the EC, across the Common External Tariff wall. Those opposed to furthering the interests of the Japanese companies and hence the resuscitation of car production within the UK via their activities, put forward varied arguments, some measured and seemingly reasonable, others less so, to support their position. This variety is itself indicative of differences in national Government positions and in the policies of car producers in Europe in seeking to respond to the Japanese challenge.

Nevertheless, from these differing, and sometimes conflicting, sources of opposition to Japanese investment, a sort of consensus began to emerge between the EC and Japan. In June 1991 the substance of an agreement, "Elements of Consensus", between them was revealed. An earlier projection by the EC Commission indicated that the Japanese share of the market by 1998 would be 18.7 per cent, of which 8.7 per cent would be imports from outside the EC and 10 per cent "transplant production" within it (Table 4.7). It also suggested that non-Japanese car plants within the UK would bear the brunt of the competitive consequences of Japanese production; the projected Japanese share of the UK market was 29 per cent, with 22 per cent accounted for by production from the UK plants of Honda, Nissan and Toyota. The new deal apparently eased the overall position, especially for producers with plants outside the UK. It reduced Japan's share of the market to 16.1 per cent, primarily because of a reduction in "transplant production" from 1,500,000 to 1,200,000 by 1999. This implies less exports from Japanese plants in the UK to other EC countries but no easing of the competitive pressures on non-Japanese plants within the UK. However, Mr. Yukata Kume of Nissan very quickly stressed that the figure for "transplant production" was the EC's unilateral internal assumption.This suggested the agreement could more accurately be described as "Elements of Disagreement" rather than as "Elements of Consensus", with a worrying ambiguity on crucial points.

Whilst new Japanese car plants brought the promise of economic revival to parts of the UK, they also threatened economic decline and social problems elsewhere as other car plants closed or reduced the scale of their operations (see, for example, Oxford Motor Industry Research Project, 1990). Shortly before the EC-Japan agreement was announced in 1991, Carole Tongue, deputy leader of the European Parliamentary Labour Party, prepared a report for the European Parliament predicting the closure of 10 to 12 car plants, with the loss of between 50,000 and 100,000 jobs in the 1990s as a direct consequence of Japanese competition (see House of Lords Select Committee on the European Communities, 1990). Clearly, which companies, plants and localities within the UK and EC would suffer as a result of this restructuring would become a matter of acute political concern, with these locationally-concentrated costs

Table 4.7 Japanese impact on the liberalised European Community market for automobiles and light commercial vehicles[1]

Market	1989	1998[2]	1998 per cent share	1999[3]
FRANCE	2,667,000	2,850,000	100.0	2,850,000
Japanese imports		162,000	5.7	150,000
Japanese transplants		155,000	5.4	
Total Japanese share	80,010	317,000	11.1	
ITALY	2,519,000	2,600,000	100.0	2,600,000
Japanese imports		148,000	5.7	138,000
Japanese transplants		142,000	5.4	
Total Japanese share	50,380	290,000	11.1	
SPAIN	1,376,000	1,475,000	100.0	1,475,000
Japanese imports		84,000	5.7	79,000
Japanese transplants		196,000	13.3	
Total Japanese share	41,280	280,000	19.0	
PORTUGAL	252,000	275,000	100.0	275,000
Japanese imports		23,000	8.4	23,000
Japanese transplants		35,000	12.7	
Total Japanese share	35,280	58,000	21.0	
UK	2,600,000	2,700,000	100.0	2,700,000
Japanese imports		189,000	7.0	
Japanese transplants		594,000	22.0	
Total Japanese share	286,000	783,000	29.0	
RESTRICTED MARKETS	9,414,000	9,900,000	100.0	
Japanese imports		608,000	6.1	
Japanese transplants		1,122,000	11.3	
Total Japanese share	517,770	1,728,000	17.5	
UNRESTRICTED MARKETS	4,531,000	5,200,000	100.0	
Japanese imports		714,000	13.7	
Japanese transplants		378,000	7.3	
Total Japanese share	788,394	1,092,000	21.0	
TOTAL EC (inc. eastern Germany)	13,946,000	15,100,000	100.0	15,100,000
Japanese imports		1,320,000	8.7	1,230,000
Japanese transplants		1,500,000	10.0	1,200,000[4]
Total Japanese share	1,310,924	2,820,000	17.7	2,430,000

[1] includes light commercial vehicles under 5 tonnes gross vehicle weight

[2] Internal European Commission projection September 1990

[3] EC/Japanese forecasts as basis for monitoring system, July 1991

[4] EC internal assumption, July 1991. Japan insists there is no ceiling

Source: European Commission internal documents (cited in Financial Times, 26 September, 1991)

counterbalancing the locationally-concentrated gains resulting from Honda, Nissan and Toyota investing in the UK.

4.8 The Local and Regional Development Implications of Japanese Inward Investment: Just-in-Time and In One Place ?

Hopes of economic regeneration centred around new Japanese car production plants have been a very important reason for seeking to attract them to areas of high unemployment. Given the plethora of such areas within the EC, the competition between them for inward investment by Japanese car companies has indeed been fierce. Nissan, for example, was lobbied by 53 locations in the UK alone before it finally chose Sunderland. Certainly, prior to the 1989 EC regulations, substantial financial incentives were used to entice Japanese companies to particular areas. Although subsidy on the scale originally received by Nissan - over £100 million of the £390 million investment for Phases I and II of its Sunderland plant - subsequently became much more difficult to obtain, this deterred neither other companies, concerned with long-term market share rather than short-term cost reductions, nor those seeking to attract them. In areas blighted by high unemployment, the relatively large numbers of jobs provided in the new factories of Honda, Nissan and Toyota were seen as providing much needed growth in labour demand. Certainly there was great selectivity as to who got these jobs, and in one sense they did little more than scratch the surface of problems of localised mass unemployment. However, the jobs created in these plants were typically the only major source of employment creation in such areas, dwarfing the net job creation effects of policies to encourage new small firms (see, for example, Hudson and Sadler, 1991).

Moreover, there were strong hopes, if not beliefs, that powerful multiplier and spin-off effects would lead to further local employment growth in component and supplier companies. To some extent, false hopes were raised because of erroneous analogies drawn by commentators such as Mair (1991) between the impacts of new Japanese car plants in the USA and those to be expected in the UK and EC. In the USA, over 300 new Japanese car component plants did indeed cluster within a few hours driving time of the car assembly plants but this was more a matter of necessity than choice, reflecting the pattern of linkages that had evolved between indigenous car production and component companies there. In western Europe, a very different pattern of such relationships existed. Thus there was never any serious possibility of dense agglomerations of new component companies locating near to new assembly plants, as Honda, Nissan and Toyota emphasised from the outset in stressing that they would seek to source components from within the EC.

This did not, however, prevent some from raising expectations as to the emergence of new just-in-time and in one place car production complexes in areas such as north east England. By 1990, however, only six such Japanese component companies had located in the vicinity of Nissan's Sunderland factory. Moreover, there is no need to invoke just-in-time as an organisational principle in order to explain this. For they specialise in producing relatively low value added and bulky components, such as plastic mouldings, trims, seat covers and seats. Such component production has always located near to assembly plants in response to transport cost considerations. Despite such evidence, there are those who stubbornly cling to notions of just-in-time and in one place.

For others, however, there has been a significant shift from assertions about the advantages of spatially agglomerated just-in-time complexes and towards those of attracting branch plant investment by car component companies. North east England, for example, is promoted as a region of national, even European, significance with 120 component producers. Less than 20 per cent of these supply Nissan but they do supply car producers in other locations. For instance, three of them are amongst the 138 component suppliers chosen by Honda for its Swindon plant.

This emphasis on the region as a car components producer reached new heights in 1991, with the release of a consultants report commissioned by the Tyne and Wear Development Corporation. As Nissan expands production at Sunderland, so its attraction to component companies undoubtedly increases. But it is important to keep this in perspective. Three points are relevant. First, there were already 83 motor vehicle component companies established in the north east before Nissan announced that it would locate there in 1984. Second, Nissan has invested heavily in developing links with its chosen set of European component suppliers, largely (about 85 per cent) located outside the north east of England. Increasing demand from Sunderland will largely be met from their existing production locations. This links to the third point. Even if Nissan increases annual output to 400,000 by the end of the 1990s while UK output rises to around 2,000,000, 80 per cent of these cars will still be produced within the middle of England. An area bounded by Merseyside, Derby, Dagenham and Swindon will continue to contain the bulk of car production in the UK and constitute a much greater attraction to such component companies as choose to relocate than will Nissan's plant, in a relatively isolated location in north east England.

Nevertheless, there may be another sense in which Nissan's location in the north east, and in like manner Honda's and Toyota's in their chosen locations, may have an impact on the local labour market and economic development prospects, associated with further inward investment in industries outside of the automotive sector. Nissan has been anxious publicly to stress its commitment to the north east whilst its decision to invest at Sunderland has been endowed with a broader symbolic significance. There is no doubt that there has been a surge in inward investment from Japan, and the Far East more generally, following Nissan's decision to establish itself at Sunderland (see Hudson, 1991). It is, however, rather more difficult to establish whether this is causally dependant upon Nissan's decision.

Any evaluation of the regenerative effects of Japanese inward investment in the north east economy, whether in automobiles or other sectors, must include consideration of the processes that initially led the region to become attractive to such companies. The destruction of much of the former industrial base in coal, steel, shipbuilding and heavy engineering, the subsequent creation of mass unemployment with all that this implied, owed much to the direct impacts of central Government policies (Hudson, 1989). It is therefore, important to remember why problems of economic decline and social stress exist in the north east in considering the extent to which they have been ameliorated as a result of inward investment from Japan (and indeed elsewhere). A similar point can be made in relation to investments by other Japanese car companies in other localities (for example, see Derbyshire County Council, 1990).

4.9 Concluding Comments

The arrival of the new Japanese car assembly plants brought a dramatic change in the geography of car production within the UK that is decisively linked to the changing international division of labour in the automobile industry. EC trade policies, the lure of the Single European Market and the fiercely competitive strategies of Japanese car companies, struggling for market share within the world's largest car market, have interacted to entice these companies to invest within the EC. The specific attractions of parts of the UK, both politically and economically, as production locations have led to substantial new fixed capital investment by Honda, Nissan and Toyota. This has created new car production spaces within the UK. Located well away from established centres of car production, in areas of high unemployment and against a national background of a significant decline in trades union membership and influence, the Japanese companies have been able very carefully to select their workforces in terms of age, physical condition, manual abilities and appropriate psychological traits of commitment to the company. On this basis, they have been able radically to redefine industrial relations, the terms and conditions of work and working practices, and achieve levels of labour productivity far in excess of those of longer-established non-Japanese plants in the UK and the rest of the EC. There has been a flexibility in accommodating local cultural specificities but equally an absence of commitments of jobs for life; this is a flexible adaption rather than a mechanical translation of practices from Japan.

It has nonetheless greatly increased competitive pressures within the EC, which will continue to be felt most acutely within the UK. Rumours of Rover switching part of its production to Russia may have been "hopelessly premature", as a company spokesman put it in January 1992, but it is nonetheless indicative of the sorts of strategic response that companies may consider to cope with growing Japanese competition (see Swain, 1992).

An important consequence of the arrival of the Japanese assembly companies has been an accelerating restructuring of the car component sector within Europe. The influx of Japanese component companies into North America, following investment in new car assembly plants, is unlikely to be repeated within the EC. There were quite different institutional arrangements between assembly companies and component suppliers in Europe, Japan and the USA. In the USA a much greater proportion of component production was internalised within the major assembly companies. In Europe the component sector was characterised by numerous small companies, often fiercely price competitive and having antagonistic relations with the assembly companies. The structure of the component sector in the UK and Europe was, however, already changing prior to the wave of Japanese investment in assembly plants as non-Japanese companies producing cars there sought to come to terms with Japanese competition on the world market. An embryonic hierarchical structure, at the time a pale shadow of that found in Japan, was beginning to emerge within the component sector. The Japanese assembly companies sought to both plug into and further transform this indigenous component sector. Their demands for total quality control and reliability accelerated the emergence of a small number of large component companies, a first tier of suppliers, typically multinational in their operations. For such companies, able and prepared to commit resources to meet the *really* significant demands of the Japanese JIT production philosophy - that is, in relation to R and D, quality and reliability - there was the reward of stable, long-term contracts with assembly companies, both Japanese and increasingly non-Japanese. First tier component companies have also begun to demand similar commitments to R and D, quality and reliability from their suppliers. Relationships of antagonistic competition have been replaced by those of competitive cooperation. Such relationships remain profoundly asymmetrical in terms of power and are not unconditionally permanent but dependent upon component companies' performance. As long as performance targets are met, however, the Japanese assembly companies will adapt to local production methods rather than impose a rigid model based on their accumulated experiences of production in Japan.

Not only have the new Japanese companies had a catalytic effect in accelerating upstream restructuring in the component sector, but they are also increasing pressures for a downstream restructuring of distribution and after sales service. Nissan and Toyota have established their own distribution networks so as to guarantee control of overall sales and marketing strategies. This is again reinforcing an existing tendency amongst non-Japanese car companies to extend control over key parts of their wholesale operations, especially in the main markets of western Europe.

Inward investment from Japan has therefore been seen as a mechanism for resuscitating car production in the UK, and more generally transforming the culture of production and industrial relations within UK industry. In addition, it has also been seen as having a very important local and regional development role. There has been a fierce competition between and within countries to secure such new investment. In part, hopes were raised because of a failure to appreciate the specific reasons as to why clusters of new component plants sprang up around Japanese assembly plants in North America. This raised, for some, the prospect of recreating regional economies around new car production complexes: just-in-time and in one place. But such optimism was badly misplaced, since there is no necessary correlation between just-in-time and in one place. Growing recognition of this led to more emphasis upon the employment creation effects of the assembly plants and of large branch plants of major component companies. These have often provided new jobs in areas blighted by deindustrialization and high unemployment on a scale far beyond that created as a result of policies to encourage new small firms there (see Hudson, 1992b). Clearly, employment in these factories revived old fears of the dangers of creating a new branch plant economy. Furthermore, there were also doubts as to the type of jobs offered in them, and concern as to who is considered eligible to fill them. But it is also true that, in terms of job creation, such large plants have a seductive attraction whilst the redefinition of global-local relations within corporate production strategies holds out the promise of more than just routine assembly plants in more locally embedded factories. Whilst Nissan may be in the process of embedding itself within the EC, and indeed the UK, however, it is important to bear in mind that its R and D and European HQ functions are mainly not located in Sunderland. Equally, however, the presence of such new Japanese plants implies job loss and plant closure elsewhere in car producing regions of the UK and EC. Such negative effects in other places have to be balanced against the positive ones for some people in areas such as Derby, Sunderland and Swindon.

Chapter 5
The German Automobile Production System Going European

Eike W. Schamp
Institute of Economic and Social Geography, Johann Wolfgang Goethe University, Frankfurt am Main, Germany

5.1 Introduction

For several decades the automobile industry was a key sector in the German economy, substantially determining its rate of growth. More than 20 per cent of domestic value added stemmed from this and related sectors at the beginning of the 1990s. As will be verified many times in this chapter, the automobile production system historically has emerged as a national system which still matters today. This above all else justifies dealing with a particular 'German' case. This production system has been extremely successful during the 1980s both on domestic and foreign markets, compared to its European neighbours, mainly a result of qualitative rather than quantitative growth due to shifts into higher valued automobiles. In what follows, some of the factors determining the growth and international competitiveness of the German automobile production system will be considered. The chapter will use Porter's (1990) diamond of international competitiveness as a starting point as it stresses the combined effects of different factors for its production and reproduction. These factors are: first, the challenging demand on the domestic market, both by number of customers as the German market by far is the largest in Europe and by clients' preferences as customers were prepared to buy larger, better equipped and higher valued automobiles. Second, the long-term competition among five large independent firms in the automobile sector, mainly sustained by technical competences. Third, the huge cluster of firms in related sectors which are highly innovative. And, finally, the kind of production factors, particularly the form of labour relations which are so determined by one strong trade union, IG Metall, its co-management strategies, sophisticated state regulation of the labour market and particular qualification systems.

Porter mainly emphasized geographical concentration as a necessary prerequisite for the combined effects of these factors. He takes, as it seems, a rather broad American perspective in dealing with distances, however. Ranges of 250 km making up a 'regional' cluster of firms in the USA will often establish national ones in Europe. Hence, it may be assumed that the particular geographical concentration of the (German) automobile sector corresponds to the approach of a national production system. The aim of this chapter will be twofold: first, to describe in more detail the different 'edges' of Porter's diamond on the national scale in order to explain the growth and restructuring of the automobile production system in Germany during the last decade. As past strategies and experiences as well as long-term investments largely determine present decisions, an evolutionary approach will be considered necessary, stressing both structural changes in the course of time and particular events. On a more general level, it will be argued that competitiveness mainly was due to technological competence being reproduced by manufacturers as well as suppliers and the workforce. But already during the 1980s the German automobile production system was gradually growing and extending into the European Community. It has to be questioned, secondly, whether the acceleration of this movement to Europe in the course of the current crisis of the sector will further erode geographical concentration and, hence, affect the long-term competitiveness of the German automobile production system.

5.2 Market Structures During the 1980s

Although the production system may be considered national, the core market need not necessarily be restricted to this area. In fact, German automobile manufacturers are prime sellers in neighbouring countries as well, such as VW in Austria or Opel in Switzerland, Belgium and the Netherlands (Table 5.1). During the 1980s, they increasingly tried to sell 'European'. Exports grew to a considerable extent, mainly to Europe and particularly to Spain (eight-fold in 1991 compared to 1981). Conversely, exports to other continents and particularly to the USA and Japan proved to be difficult, partly due to fluctuating currency exchange rates. For this reason, some scholars speak of a retreat to Europe during the 1980s.

The domestic market remained the main economic base for German automobile manufacturers, though, as is clearly shown exemplified by VW (Table 5.1). Given the number of 489 automobiles per 1,000 inhabitants (VDA 1992) and more than 3 million cars sold in 1993 Germany is the largest European market. Although it has been considered as a quasi open market for a long time (Bloomfield, 1978), which is exceptional amongst the larger

Table 5.1 Stock of sold VW cars in different European countries, 1990.

Nation	Stock in millions cars	Percentage of national stock
FRG (West)	7.9	26
Italy	1.5	6
UK	1.3	6
France	1.2	5
Austria	0.7	23
Netherlands	0.6	10
Belgium	0.5	12
Switzerland	0.4	14
Spain	0.4	4

Sources: VDA 1991

European domestic markets, the German market has a quite particular feature. An extended institutional framework has been brought into existence favouring the private use of automobiles, via collaboration between the state (for example, by huge investments into construction of free entry highways and tax reliefs for private car commuters), large German manufacturers (joined in the Association of Automobile Industry, VDA) and the user's club ADAC, Europe's largest and probably most powerful automobile club. This corporate system of creating and safeguarding markets deserves closer analysis than can be given here. Suffice to say that it works both for open markets for domestic producers in Europe and cautious protection against Japanese competitors.

For many years, markets were segmented vertically, that is by size of cars and cubic capacity of engines. Although German automobile manufacturers now only hold a minority portion of 45 per cent in the market segment of small cars - which, furthermore, mainly are produced outside Germany - they still maintain dominance in the other segments of the automobile market (Table 5.2). In no other European country is the portion of the upper segments as high as in Germany (Diez, 1988). Hence, even if one could roughly differentiate German automobile manufacturers between three domestic volume producers (Ford, Opel/GM, VW) and three specialised producers (Audi/VW, BMW, Mercedes-Benz), all firms simultaneously tried to up-grade their products. This has been a major change in strategies after the crises of the early and late 1970s in which the former low cost mass production strategies have been seriously hit. Clients increasingly expected special equipment, such as sun roofs, servo steering, thermal insulation windows, centralised locking systems, anti-block systems and the airbag. In consequence, competition among German automobile manufacturers focused more on technical innovativeness than on price. Hence, technical competence is one of the main features reproduced by the characteristics of the German market.

Table 5.2 Market portions of automobile producers in Germany according to different market segments, 1991.

Size of cars	Models	Percentage of total German market	Percentage of automobile firms producing in Germany
small cars	Fiesta, Corsa, Polo	20	45
lower middle class	Golf, Kadett Escort	31	68
middle class	BMW 3, Vectra Sierra, Passat DB 190, Audi	36	80
upper middle and upper class	DB different models, BMW 5,7,8 Senator, Porsche Audi different models	13	88

Sources: Frankfurter Allgemeine Zeitung 45, 19 March 1992.

During recent decades, technical innovations in the two upper market segments mainly determined international competitiveness even for products in the lower segments. An example of this trend is the electronic injection pump which clearly showed a vertical diffusion from luxury cars to lower middle class cars. The trickle-down process could be explained by greater acceptance of technical innovations by wealthier customers, by the cautious strategies of automobile firms in offering technical innovations initially as special equipment in small series luxury cars, by the changing consumer preferences as real incomes rise and, in total, by the emphasis placed upon the technical characteristics of automobiles by German customers (Diez, 1988). These are, as Diez (1993) points out, mainly visible characteristics of automobiles.

What has not been fully foreseen, however, by German automobile manufacturers was the shift in consumer preferences from a vertical segmentation of markets to a horizontal one. Market shares of saloon cars decreased from 86 per cent in 1982 to 74.1 per cent in 1992 while, for instance, that of estate cars rose to 14.6 per cent (from 8.2 per cent), convertibles to 2.3 per cent (from 0.5 per cent), off-road cars to 2.4 per cent (from 0.9 per cent) while a completely new type of car emerged, known as multi-purpose vehicles (1 per cent). Import shares were rather large in these segments, for example 20.2 per cent for convertibles, 26.5 per cent for coupés, 97.6 per cent for off-roads and 100 per cent for people carriers in 1992 (Diez, 1993). For the time being, an additional increase in the demand for small cars ('city cars') is expected, such as the Renault Twingo or Fiat Cinquecento.

This has been supplemented by an emerging challenge to domestic production in Germany through the Japanese entering the market segment of upper middle and luxury cars via imports from outside the EC rather than from transplant production (see Hudson, this volume). During the 1980s, Japanese penetration of lower market segments mainly occurred at the expense of imports from the European competitors Renault and Fiat. As a result, the percentage of Japanese registrations grew relatively slowly from 9.5 per cent in 1981 to 14.1 per cent in 1993, including the east German market. Whereas for many years the Japanese have been considered weak in the upper middle and luxury class market segments, Dolata (1990) assumed that with the emergence of transplants in the USA during the 1980s they will be able to learn from innovative competition on that market, the world's largest market for upper middle and luxury class cars. Additionally, however, the large three American firms have fundamentally restructured and become competitive in the early 1990s, particularly with the rising yen. Hence, the German market in this segment which, until now, has been mainly held by Mercedes-Benz, BMW and Audi increasingly is coming under pressure.

During the last decade, automobile manufacturers have made considerable efforts to strengthen domestic distribution systems. They have increasingly created guaranteed markets, not least through tax free sales of new cars to their own employees who get a reduction of 15 to 23 per cent on final price and, hence, are able to make extra profits in reselling the used car one year later. This already amounts to 7 per cent of total annual sales. A further 20 to 30 per cent of sales, mainly of the volume producers VW, Opel and Ford, were at a discount to large clients and authorities. Furthermore, most producers have established close contacts to car rental and leasing firms, 65 per cent of car leasing enterprises currently belong to automobile producers, VW being the largest with a market share of 23 per cent in 1990. Private car leasing has seen a boom during the 1980s (Schamp, 1993). Leasing firms, however, cause a squeeze on the prices of used cars as they resell cars within a year of initial purchase. In the end, however, these protected market segments were only able to slow down the distribution crisis of the early 1990s. Considerable efforts have also been made in the reorganization of distribution channels, either via a strict reduction and selection of the car dealer's network - the distribution system in Germany being organized according to the brand principle - or via technical means such as closed telecommunication networks from the manufacturer to the dealers' network and via a system of specialized repair shops. The general shift in market segments, however, will force the automobile manufacturers completely to reorganize distribution once more in the near future.

5.3 Competitiveness via the Technological Competences of Automobile Producers

Automobile producers shaped their strategies on the German and European markets according to their specific resources and history. Consequently, the structure of the main six producers is very heterogenous. While there are three companies which produce almost exclusively in Germany (Mercedes-Benz, BMW, Audi/VW), three global companies, that is the German VW as well as the two American companies Opel/GM and Ford, have a long-established European production network. Two of those are young, established just before and after the Second World War (VW 1938, BMW 1952). Three of them are among the six volume producers existing in Europe, VW being the largest on the European markets (Table 1.3). Mercedes-Benz and BMW are considered to be low volume specialised producers but have each had an annual production of about 500,000 cars for many years.

5.3.1 The emergent spatial pattern

The spatial structure of automobile production in Germany emerged in different stages after the Second World War, during four decades of almost continuous growth but strong competition. The reconstruction phase is characterized by the entrance of a lot of newcomers producing a diversity of small cars. It ended in the 1960s, with concentration in a small number of companies each producing a diversified range of cars (Berg, 1984). During the concentration process some establishments of firms which became bankrupt were taken over by surviving companies, such as the large Dingolfing plants of BMW taken over from Glas or the Audi plants in Neckarsulm and Ingolstadt taken over by VW. As a result, the spatial pattern of car production is mainly characterized by location in agglomeration areas where the main integrated plants of most companies are located as, for example, Mercedes-Benz in Untertürkheim and Sindelfingen near Stuttgart, Opel in Rüsselsheim near Frankfurt am Main, Ford in Cologne and BMW in Munich (Fig. 5.1). An exception to the general trend is the state-planned VW plants at Wolfsburg, where what is currently the world's largest single automobile plant (60,000 employees) has developed.

In the 1960s, which could be described as the extensive expansion phase, surviving companies reached the limits of their production capacities, particularly in urban agglomerations, and started to establish greenfield assembly plants and component plants separately, VW creating a full system of branch plants in Lower Saxony (Kassel 1960, Emden 1964, Salzgitter 1970), Opel/GM going to Bochum (1964) in the Ruhr area and Ford to Saarlouis (1968) on the fringes of the old industrialised Saar region. Obviously firms tried to establish firm-specific regional production clusters, particularly BMW

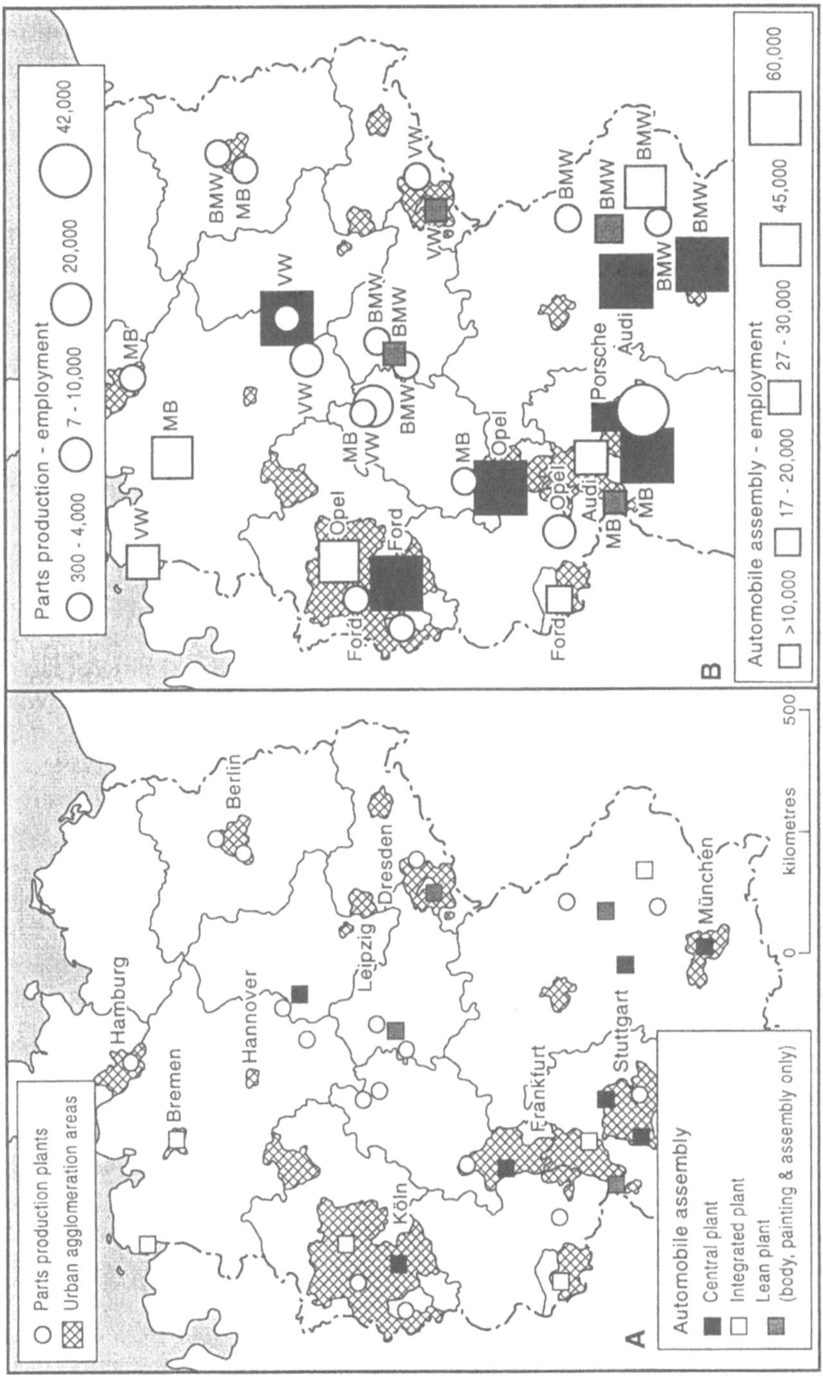

Fig. 5.1 The spatial pattern of automobile production in Germany as of 1992.

in Bavaria and VW in Lower Saxony. Apart from the more familiar locational factors such as accessibility, availability of space and low land prices, this has been principally caused by both the availability of a qualified industrial workforce in times of full employment and the corporate interplay of the different states (*Länder*) and firms locating in what was considered to be a growth sector in problem regions. This held true for old industrialised areas and towns in crisis at that time such as the Ruhr area with coal-mining and steel production, Emden (and later Bremen) with its shipbuilding industry, Kassel with railway engine construction and Salzgitter with iron ore mining. Smaller plants for parts production, however, were allocated to rural towns such as Düren (Ford, 1968), Kaiserslautern (Opel/GM, 1966) and Landshut (BMW, 1967). Thus most integrated quasi one-plant companies developed into more or less regional large multi-plant companies, able to implement new locational strategies in the internal division of labour in the future decades.

5.3.2 Choices in product and process

As noted earlier, the competitiveness of German automobile producers was largely due to technical competence stemming from the production of upper middle and luxury class cars and to the image which has spread from these to cars in other segments of the automobile market. But the rough dichotomy of an upper and lower segment in automobile production began to be eroded during the 1960s. Automobile producers subsequently have tried to cross the traditional boundaries of market segments and have diversified, mostly vertically. VW, having grown up with only one make, the "Beetle", started production of a middle class automobile in 1961 and took over the middle class producer Audi from Mercedes-Benz in 1965. Audi developed into an upper middle and luxury car producer during the 1980s. BMW started middle class automobile production in the 1960s and grew to be the main competitor to Mercedes-Benz in the luxury automobile segment during the 1980s. Conversely, Mercedes-Benz started production of an upper middle class car (DB 190, new version) in the early 1980s to compete with BMW and Audi. Thus, the ideal type segmentation of volume and low volume specialised car producers in Germany is not only arbitrary but fails to reflect the ongoing competition between domestic producers, enforcing continuous technical innovations both in products and production methods. With the appearance of Japanese competitors, however, this competitive situation was further intensified.

The tendency of German automobile producers to look for technical solutions became very obvious during the long decade of crises from 1973 to the early 1980s. Adjustment strategies led to an intensification of production in two ways. First, no plant was closed (although this was considered) but the workforce was reduced, sometimes drastically, at existing plants. VW, for example, reduced its workforce between 1971 to 1975 by nearly 29 per cent, although largely avoiding mass dismissals. But reduction via various other means, which have resulted from discussion with the trade union IG Metall, resulted in an increase in labour qualification, as it has been very selective: principally foreign workers (so-called guestworkers), unskilled and female workers in parts production have been removed, thus sparing qualified male workers in assembly and white-collar employees (Lompe et al., 1991). Second, improved qualification enabled most automobile producers to start an enormous process of modernising production equipment in the late 1970s, particularly via investing in computer supported machinery. Computer-assisted automation first took place in stamping, in body-making and parts production. VW, for example, started its own production of robots, particularly for body welding. Most assemblers looked to gain further competence in running computer assisted machinery by integrating software departments or buying software firms. The consequence of this modernisation process was further labour shedding, which has been masked by growing production capacities during the 1980s, however. As a result, German automobile production now is much more capital intensive compared to its Japanese competitors (Weißbach and Weißbach, 1987). But producers were looking for different paths of technical improvement, VW mainly concentrating on the further automation of assembly in the early 1980s, Opel/GM and Ford establishing a computerized network between their different (European) plants while BMW remained very cautious in introducing computerized manufacturing (Dolata, 1990). This may be due to persisting differences in production methods between high volume producers of lower middle class cars (such as the VW Golf) and low volume producers of upper middle class cars (such as BMW series 3). In fact, the strong automation strategy in car production lead to rather rigid production methods, that is to inflexibility, which necessitated a high level of capacity utilization, and this in turn may have been one reason for long-term low profitability of VW.

5.3.3 Two joint strategies: Europeanisation and flexibilisation

The particular technological trajectories of automobile producers in Germany were not sufficient to cope with growing but enormously diversifying markets during the 1980s. Assemblers therefore chose two different strategies to meet new market requirements: first, the strategy of Europeanizing production via establishing a more sophisticated internal division of labour between European branch plants and, second, strategies of increasing internal and external flexibility of production. To this point in time, both strategies have been so successful that assemblers in Germany only marginally lost market shares to their Japanese competitors. In total, domestic market shares of assemblers located in Germany diminished from 75 per cent in 1981 to 67 per cent in 1991 but, measured in absolute terms, domestic production rose from 3.28 million automobiles in 1981 to 4 million in 1991 (VDA, 1992).

In considering the Europeanisation strategy, it must be stressed that volume assemblers basically continued a strategy during the 1980s which had already been successful during the expansion phase in the 1960s. At that time assembly of small and lower middle class cars was relocated from parent plants to new greenfield sites, such as the Kadett to Bochum (Opel/GM), the Escort to Saarlouis (Ford) or the Beetle to Emden (VW). This enabled firms to increase production of higher class cars in core plants. Some companies had already experience with European SKD and CKD assembly of automobiles, such as Opel/GM in Antwerp since 1953/67 and VW in Brussels since 1953 and now started to relocate integrated production. In the early 1980s, the new industrial policy of Spain and Portugal and, later, the enlargement of the European Community opened up those countries as a strategic option for producers to establish a European production network. First, American firms went European, when Ford reorganised to Ford Europe in 1967 and relocated production of one of its small models (the Fiesta) to Valencia in 1976. Opel/GM followed with a greenfield plant in Zaragoza to produce the Corsa in 1982 (Bloomfield, 1991; Dicken, 1992a). They established a European production network in its proper sense, both by parallel production of models at different European sites (such as the Fiesta at Valencia, Dagenham and Cologne) and cross-plant exchange of parts (Dicken, 1992b; Wells & Rawlinson, 1992). VW was a late-comer when it took over the Spanish company Seat with sites in Barcelona (mainly producing the Ibiza model) and Pamplona in 1985. And it pursued a different strategy in dedicating one model to one plant and keeping Seat as a separate make of small cars. This was continued in the early 1990s, when VW took over the Czech Skoda automobile company, with the aim of penetrating the east European markets with small cars. Basically, however, all three producers created a segmentation between foreign production sites in the European fringe, producing mainly small cars, and domestic production, while retaining capacities and competence for R&D and production of larger, more expensive models in Germany.

Already during the 1970s Opel consistently undermined integrated production by relocating the assembly of small and middle class cars to foreign sites and by creating a subsidiary company responsible for components production, the GM Automotive Components Group with its headquarters in Paris and production sites in France, Spain and the UK. Shifts of production within the company are now easy as automobiles, in principle, are produced in parallel at different sites. Thus new investments in greenfield sites such as Eisenach in eastern Germany or Szentgotthardt in Hungary will enable a further shift of middle class car production from west Germany. Ford, however, even kept production of the small Fiesta at its Cologne plants, producing in parallel, probably due to long-term productivity differences among its different European plants (see Hudson, this volume, Wells & Rawlinson, 1992).

Furthermore, competences have been reinforced in Cologne as Ford extended production of engines and gearboxes, inter alia for the multi-purpose vehicle jointly produced with VW in Setubal in Portugal as from 1994 (see Ferrao, and Vale, this volume).

The second strategy of seeking enhanced competitiveness has gained important impetus as well, particularly in the context of the recent debate on lean production (Womack et al., 1990). It is now almost common knowledge that productivity gains are no longer to be achieved via rigid automation, that is through technical changes alone, but with changes in production oriented towards both a greater internal and external flexibility. Flexibility is an ambiguous term and its meaning should be clearly defined but this is not the place to discuss the notion of flexibility in more detail (see, for example, Gertler, 1988; Coriat, 1990). Internal flexibility, that is production in small batches but with more variants which reduces minimum necessary capacity, will be achieved by both using new flexible machinery (but to a reduced extent) and by restructuring labour processes. External flexibility is mainly sought in vertical disintegration by out-sourcing.

Automobile producers in Germany have made changes in the internal operation of production very hesitantly. Production organisation at core plants has not been radically altered and flexible production has only been introduced with new car models at branch plants. With rigid automation, changes in machinery could only be carried out simultaneously with changes of models. Therefore, in former times best practice technology was allocated to parent plants, along with production of the principal car model of the firm. In a transition period from rigid (Fordist) to flexible (post-Fordist) organisation during the mid-1980s flexible production, together with new models, was first introduced at two greenfield sites (Mercedes-Benz in Bremen and BMW in Regensburg) and one completely refurbished site (VW in Emden), but successful 'bread and butter' models still were produced at the core plants. This enabled companies to go through a cautious learning process with new flexible technologies (Schamp, 1991).

5.3.4 The 1990s

In contrast to most European competitors, the high demand for automobiles produced in Germany seemed to intensify with the emerging new east European markets. Expecting further growth, firms increased capacities in the early 1990s. Experiences with flexible production were now drawn upon extensively in making investments in new plants and even in making changes in old core plants, as was exemplified with the flexible production of the new Golf III model in Wolfsburg. Opel at Eisenach in Thuringia (for production of the Vectra and Astra), VW at Mosel in Saxony (for a parallel Golf production) and Mercedes-Benz at Rastatt established 'lean' plants equipped with flexible machinery and using new forms of labour organisation. In contrast to Nissan in the UK, or even their own former location policy during the 1960s, automobile producers now only invested in modern flexible production sites in which a labour force highly qualified in automobile production was available. Both Eisenach (the Wartburg model) and Mosel (the Trabant model) have been the main private car production centres in the former GDR whereas Mercedes-Benz drew upon skilled labour from a neighbouring truck plant in Rastatt. Most radically, principles of lean production have been introduced encompassing new forms of labour organisation (team work), flexible machinery and a sophisticated modular sourcing based on systems assembly in neighbouring suppliers' plants. Whereas older plants have a net product of up to 47 per cent, new plants only have a level of 30 per cent. At Eisenach and Mosel, this is partly a result of the absence of body stamping. Stampings are delivered from Wolfsburg to Mosel and from Zaragoza to Eisenach by train. Employment creation, which was already planned to be rather low, will be further reduced by the current crisis to approximately 2,400 jobs in Mosel, 5,000 in Rastatt and 1,800 in Eisenach. Two points have to be stressed in considering the significance of these new plants for the restructuring of the automobile production in Germany: first, for the first time core products of automobile assemblers will be produced in a flexible way which, secondly, makes relocation of production from old core plants in urban agglomerations to new places, characterized by lower wages, availability of space and new machinery, much easier than before. The old and restricted plants of Opel/GM in Rüsselsheim and BMW in Munich may be particularly vulnerable under these new circumstances.

Companies, however, overestimated demand and went into crisis, perhaps with the exception of BMW, which gained as a result of successful new models and a cautious capacity policy. In 1993, investment plans were reduced, causing serious political tensions for VW both with the Czech (in the case of Skoda) and Spanish (in the case of Seat) governments. Opel stopped extensions at Eisenach and Mercedes-Benz cancelled investments in body stamping and spraying installations in its very modern but underutilized plant at Rastatt. Older plants were hit as well, such as the Audi plants (Neckarsulm and Ingolstadt) or the Saarlouis plant of Ford. Hence, negotiations with labour (see below) and the hectic search for new models took on added urgency. In fact, VW/Audi and Mercedes-Benz recently decided to produce their new city car models at Ingolstadt and Rastatt, respectively, and, thus, to stay in Germany with this kind of vehicle.

Structural economic reasons for relocation to the European fringes, such as the highest labour costs and high currency exchange rates, still continue to exist. The emergent European division of labour will be deepened, as is shown by recent decisions over the location of companies' technical competence centres. A complex network of research laboratories of automobile producers is emerging. Supplementing R&D at the parent plants, some companies established new R&D centres, such as BMW in Munich (1986) or Mercedes-

Benz in the scientific town of Ulm (1988). Companies with an European production network reinforced R&D responsibilities at their German parent plants, such as VW for engine and chassis development of all makes (except Audi but including Seat and Skoda) at Wolfsburg or Ford at Cologne, where competences are increasing compared to the R&D centre at Dunton in the UK. Even Opel/GM has kept its European development centre at Rüsselsheim after creating a new European headquarters at Zürich in Switzerland (Lompe et al., 1991). This European division of labour may well be strengthened further via the expected concentration process of European automobile producers. While the failed Renault-Volvo merger has not directly affected the German production system (except via the relocation of Volvo Germany to the location of Renault Germany near Cologne), the recent take-over of 80 per cent of Rover by BMW may cause a complete reorganisation of the spatial pattern and long-term strategy as a former specialised producer becomes a volume producer.

Diminished expectations on the domestic market certainly have caused the specialised producers of upper class cars to revise their strategy. Obviously, a shift from a local player/global exporter to a global player strategy is taking place. Apart from the large American (mass producing) companies, only VW of the German companies has been a global player. Particularly on the USA (but also on the Japanese) market VW has been unsuccessful and closed down its Westmoreland plant (see Krumme, 1981), while the company sets its hope on the Puebla plant in Mexico for the newly arising NAFTA market and on the Shanghai plant for China. Mercedes-Benz and BMW were national producers with considerable export shares to the USA and Japanese markets. BMW recently announced investments in a new USA plant at Spartanburg in South Carolina, commencing convertibles' production in 1995. Mercedes-Benz refurbished its former Spanish van plant at Vitoria to produce a new multi-purpose vehicle model from 1995 and announced investments into a new USA plant at Tuscaloosa in Alabama for the production of an off-road model. Hence, both companies reacted to the newly emerging horizontal segmentation of global markets and tried to occupy the higher valued parts of those segments.

5.4 Organisational and Spatial Restructuring of the Components Sector

In the current lively debate about flexible production and extended variability of products in emerging horizontal market segments, the problem of the increasing number of different parts and components has been underestimated. It is only during the current crisis that automobile companies are developing a standardisation strategy for parts in search of considerable cost reductions. For instance, VW announced a reduction in the number of different axles for the

Golf model from 30 to 7, expecting a cost reduction of 1 billion DM. Of course, this new strategy will have tremendous effects on component suppliers when standardisation affects not only different variants of one model but also different makes as well.

Recent restructuring gave rise to an extended debate about the future role of the original equipment component sector in Germany. Automobile manufacturer's strategies of lean production and relocation within Europe created new demand for components but also much pressure on component suppliers. Scholars either speak of a new partnership of assembler and supplier (for example, Meinig, 1993) or emphasize the new kind of domination by assemblers and increasingly ruinous competition among suppliers which particularly threatens the host of small and middle sized firms which is so typical of German manufacturing structures (Mendius and Wendeling-Schröder, 1991; Semlinger, 1991). Unfortunately, there are no clear-cut definition of, nor statistics on, the automobile suppliers sector as it crosses statistical sectors and market entry and exit are frequent occurrences. So, there is only anecdotal evidence. According to Lamming (quoted by de Banville and Chanaron, 1991, 51) German suppliers account for approximately 40 per cent of all suppliers in Europe, including some 450 large companies and 5,000 small and medium sized ones. Indeed, a strong base of internationally competitive suppliers to automobile producers has existed in Germany for many years. This is clearly shown by export statistics, with approximately 40 per cent of output exported in 1992, according to the VDA. Nowhere, except in Japan, is the balance of exports and imports as positive as for Germany.

The old geographical pattern of suppliers had two main features: first, some historically evolved regional clusters of specialised industries, and, second, an increasing tendency for spatial decentralisation into rural areas, particularly in southern Germany, during the 1980s. Highly specialised metal and non-metal working suppliers emerged from the traditional metalworking industries south of the Ruhr area (Bergisches and Märkisches Land). Specialised clusters - probably typical Marshallian industrial districts (Beccatini, 1991) - still survive, such as the locks industry in Velbert or drop forges in Plettenberg, in the Ruhr valley and in Solingen. While 65 per cent of the output of the latter was still destined for the automobile industry in 1991, this sector is under extreme pressure from global competition. Metalworking as well as electrical suppliers emerged in Baden-Württemberg, notably Bosch, and the Rhine-Main area around Frankfurt houses a further cluster of suppliers (Rehfeld, 1992). Traditional regional clusters in Saxony had been completely re-organised into vertically integrated combines of automobile manufacturing during the period of state socialism. Privatisation after unification destroyed such hierarchical networks and resulted in many closures (Kowalski, 1992; Lungwitz and Kreißig, 1992). An almost completely new pattern of component suppliers had to be established when Opel and VW went to Eisenach and Mosel, respectively.

Those clusters depending upon the historically evolved technical competence of many small and medium sized firms, particularly in metalworking and electrical appliances, declined in importance during recent decades, partly due to changes in the material content of automobiles. Low-tech labour intensive production increasingly has been shifted towards rural areas characterised by lower wages. New 'districts' emerged as even smaller companies were able to find new niches there, for example with the plastics industry in the Vechta district of Lower Saxony where former cork processing companies changed to plastic processing in the mid-1970s and were able to enter the automobile suppliers market. In fact, employment doubled between 1978 and 1988 to 4,620 in that district (Olberding, 1990). Rural areas along the German-German border gained as well from this relocation process. Whereas they have already been threatened by older competition from lower wage countries such as Austria or Spain, quite obviously they face a tremendous low-price competition both after the opening of east European countries and with the intensifying crisis on automobile markets.

In contrast to other countries with a strong automobile industry, component and parts suppliers in Germany have, to date, possessed considerable autonomy both in ownership and in technical knowledge. Compared to average R&D expenditures of only 3.7 per cent of annual turnover, the automobile industry suppliers spend 6.3 per cent (Meinig, 1993). Suppliers have repeatedly provided important stimuli to technical innovations in automobile production. Additionally, automobile producers until recently required a high degree of sectoral diversification from supplying companies, in order to make them less vulnerable to crises in car manufacturing. German automobile producers still buy more than 80 per cent of supplies from companies in Germany (Bochum and Meißner, 1990). New strategies of automobile producers, such as the creation of an own Automotive Components Group by Opel/GM, but more particularly tendencies towards establishing lean production combined with modular sourcing, may well drastically change the position of parts and components suppliers. While automobile producers further out-source parts production, they simultaneously radically reduce the number of supplying firms - which does not necessarily mean a reduction in the number of plants delivering to automobile assembly. Ford was a pioneer in modern quality control of supplying plants but now this is common practice among automobile firms. Consequently, the selection strategies of assemblers are generating strong competition among the parts and components suppliers.

Modular sourcing, as a key factor in restructuring of the automobile production system, requires new competences from companies in assembling components and prefabricated systems. Traditionally parts suppliers acquired knowledge in handling one or two materials but now connecting several materials is what matters. Apart from those competences in production, components suppliers must acquire new managerial abilities in steering the logistical process of their own suppliers. As a result, component suppliers no longer will be able to produce for different markets within the same plant as new competences will only be oriented towards their customers in automobile production. In addition to this competition on innovation, price competition is rising as wealthy newcomers enter the market and component suppliers try to up-grade into what has been called a first tier supplier segment. Both as a means of competition as well as a result of competition, three different trends in the parts and components supplier industry determine the current organisational and spatial restructuring: an emerging just-in-time production, concentration, and internationalisation.

When automobile producers outsource component assembly while simultaneously requiring just-in-time (JIT) delivery, the spatial consequences of such reorganisation increasingly become unclear. Early lessons from Japan, the USA and even Germany suggested a trend to new quasi-integrated production complexes (Dunford and Benko, 1991; Schamp, 1991; Mair, 1992). Even if one confines oneself to JIT in its proper sense, that is synchronized delivery of components to the final assembly of the automobile, the spatial outcome can be considered as the result of a strategic calculation of the company combined with learning and institutional innovation: apparently, the later JIT is introduced, the smaller will be the probability of spatial clustering (Lung and Mair, 1993).

An early example of JIT in Germany is seat production. This started with the new seat plant of Keiper-Recaro in Bremen in 1983 for JIT delivery to the nearby Mercedes-Benz plants and, later, to the Emden factories of VW. Increasingly, automobile producers outsource seat assembly. Seats are now delivered JIT by Schmitz to the BMW factory in Regensburg, by RHW and Naue to BMW in Munich, by Naue-Johnson to Opel in Bochum, by Keiper-Recaro to Audi in Ingolstadt and Neckarsulm and new greenfield plants are under construction at present near Opel in Rüsselsheim and Eisenach (Lear-Nosag) and near VW in Mosel (Naue-Johnson). These new plants are characterized by minor local employment effects, with a maximum of only 200 or 300 employees. They need at least two automobile assemblers in their neighbourhood and are particularly attracted to sites in small rural towns where companies are able to implement a highly flexible work organization and pay relatively low wages. Morris (1992) quite right called them satellite plants but this holds true in two respects. Satellite plants are not only deeply integrated into the process of automobile assembly but are also disintegrated outposts of large components companies fighting for market shares. Schmitz belongs to Europe's largest seat company, the French Bertrand Faure-Edepa which took over RHW in 1990, Naue was completely taken over by the American Johnson Controls Inc. in 1992, Nosag has been a subsidiary of the American Lear company since 1966 but is only now entering the seat assembly market in Germany. After its merger in 1988 Keiper-Recaro emerged as the only German family enterprise among the large seat assemblers which already has grown internationally. Satellite plants with JIT delivery, therefore, can be considered as part of the competitive strategy of multinationals assembling modules or systems.

Prerequisites for this synchronized JIT production are large volume, high value, great variability and changing demand for modules (Doleschal, 1991). Few components other than seats are currently adapted to JIT delivery and assembly in satellite plants, among them bumpers, plastic fuel tanks, door linings and instrument panels. The new VW plant at Mosel has been considered a further step in the evolution of satellite plants. For instance, Hella KG, a major producer of headlamps, has started to assemble the total 'frontend' for the new Golf model composed of bumper, radiator, radiator grilles, headlamps and ventilator. Such firms are now called systems suppliers. Large component suppliers such as Peguform for plastic parts, Bosch for electronics and Siemens for cables will, then, be second tier suppliers, although they will not lose their particular contracts with VW. Furthermore, due to poor infrastructure systems suppliers are obliged to locate within 6 km of the Mosel plant. Whether this will be the future form of local production organisation remains to be seen as investments in the VW plant have been reduced drastically. This may even raise questions as to the further existence of the total complex.

Increasingly, the meaning of JIT shifts from location to organisation. Both automobile and component companies look for an improvement in the flows of goods and learn by interaction to create new organisational forms (Lung and Mair, 1993). In consequence, there are some doubts as to whether new quasi-integrated regional production complexes will emerge more extensively (Dunford and Benko, 1991; Schamp, 1991) as JIT delivery will mainly result in reduced stock holding of half a day or a single day. This can be compatible with long distances to the customer if infrastructure is good. According to Bertram (1994) flexible assembly at the north German locations of Bremen and Emden is currently being served from traditional locations of the supplier industry in western and southern Germany. Time flexibility of delivery, therefore, will be better achieved by improvements in transport logistics than by new satellite plants. Warehouses in the neighbourhood of automobile assembly plants have emerged, managed by specialized companies in logistical services on behalf of either the automobile company or the suppliers. This even applies to steel, which now is delivered at short intervals from nearby warehouses to the Audi and Opel plants (since 1990 and 1992, respectively). Computerized information networks between logistical companies, computer-assisted management of warehouses and standardization of small containers reducing handling time from parts production to automobile production, characterize many of the organisational methods of improving the logistical chain between supplier and customer. Organisational innovations even allow increasing long distance railway transport in JIT delivery combined with polarised warehouse complexes; both is more likely to occur in future.

JIT and its further implications (such as stock reduction, improvement of quality etc.: see Gupta and Heragu, 1991) necessitates close coordination of the assembly processes of both automobile company and systems supplier. This means, for example, full access for the assembler to internal information about the supplier, even if it is a second or third tier supplier, closed telecommunication networks to each automobile assembler, co-determination of profits as the automobile company tries to get full knowledge about costs but without, in return, longer-term contracts as might have been expected (Meinig, 1993). Component assemblers face, in heightened form, the dilemma of simultaneously safeguarding flexible JIT delivery and obtaining economies of scale in production. Seats companies provide an example of how component suppliers may solve this problem more generally in future, that is through splitting production into flexible assembly near the customer and more or less rigid production of standardized parts in low wage countries on the European periphery. In an environment of simultaneously increasing innovation and price competition only financially powerful companies, which are able to achieve economies in production via an internal spatial division of labour, can succeed. Hence, an increasing concentration is taking place among components suppliers, combined with a growing Europeanisation of parts production.

Recently, foreign companies, as well as German firms pursuing a conglomerate strategy, have entered exclusively by merger into the segment of the largest components suppliers (Table 5. 3). Under the umbrella of Krupp Hoesch Automotive, for example, the division of the steel company Krupp Hoesch with more than 18,000 employees and nearly 3 billion DM annual turnover, are joint ventures Hoesch Federn (springs), Bilstein (shock absorbers), Krupp Brüninghaus (spring elements), Hundhausen (cast parts), Gerlach (drop forgings), Drauz (stampings), Camford in the UK (steering columns, brake discs, etc.), and, finally Krupp Metallurgica CMPO at Limpo in Brazil. Not only are German large component companies further concentrating, such as Bosch, Siemens, Mannesmann, and others, but at the same time they - and smaller suppliers - continue to Europeanise (Bochum and Meiner, 1990; Wortmann, 1990). According to recent statistics from the VDA, in 1993 German suppliers had 422 European subsidiaries, including 100 in Spain, 67 in France, 64 in UK, and 41 in Italy. But increasingly large and smaller suppliers go to eastern Europe as well. Consequently, Europeanisation of large components companies reflects both pressures to acquire new markets and to open up low wage production sites. The increasing relocation of low-tech production, however, does not stop at Europe's frontiers but even extends to southeast Asia, where, for example, Bosch located radio receiver production and VDO AG electronic devices production.

Whether recent investments in east Germany point to increasing capacity overall or, in the long run, to relocation from traditional west German sites, will be a matter of debate, *inter alia* among companies and trade unions. The consequences of changing market structure and a growing international division of labour among parts and components suppliers, however, matter more. At the top of various branches a relatively small oligopoly of firms has emerged, which has achieved considerable competence in modular sourcing. These firms are obliged to cooperate with automobile companies in design and,

Table 5.3 Some major takeovers in the German components industry, 1988 to 1992.

Acquiring firm	Core sector	Year of acquisition	Acquired firm	Acquired component sector
German				
Bosch GmbH	electronics	1991	Moto-meter	instruments
Krupp Hoesch Automotive GmbH	steel	1988	Bilstein	shock absorbers
Kloeckner AG	steel	1989	Benecke	interior trim
		1989	Kautex	plastic fuel tanks
		1990	Eurotec [1]	plastics
Mannesmann AG	steel	1991	Boge	shock absorbers
		1991	VDO	instruments
Foreign				
Cockerill Sambre	steel	1990	Ymos	metal parts
B.Faure-Epeda	seats	1990	RHW	seats
Johnson Controls	seats	1992	Naue	seats
Trelleborg	mixed	1989	Ellerbrock	rubber parts
Valeo	mixed	1989	Tibbe	locks
		1991	Renak	clutches

[1] with Peguform in FRG, Manducher in France and Inergo Plastica in Spain.
Sources: Automobilproduktion, Handelsblatt

hence, have recently invested in new R&D centres. An emerging complex spatial division of labour within large components firm could be imagined, with new R&D centres being decentralised to new locations in Germany, satellite plants clustering around flexible automobile assembly, high tech parts production surviving at traditional sites and low tech parts production being relocated to low wage regions in Europe or abroad. Obviously, such a spatial restructuring may well be suited to maintaining or even improving technical competences within the German supply industry. There is, however, one problem: what will happen to the numerous small and medium sized suppliers which are unable to establish such an inter-regional and international production network? Some medium sized firms already have experience in establishing foreign branch plants, generally located in nearby (increasingly east European) countries; others are cooperating in shifting production abroad. But apart from current complaints about the increasing exploitation of small suppliers, the strategic and spatial consequences are not yet well known.

5.5 Consequences for Labour

While employment in the automobile production system diminished radically in some European countries, employment in the German automobile industry and particularly in the parts and components industry continued to increase during the 1980s. This was partly due to growing capacity but more to increasing variety in equipment, the growing content of high value components in the automobile and to the successful policies of trade unions, for instance in reducing the number of weekly working-hours and pushing through the so-called humanization of work (Jürgens, 1991). But in the crisis of the early 1990s, the German automobile sector obviously had to catch up in restructuring. Some large companies have already cut employment, such as Mercedes-Benz by 25,000, VW 13,000, and Opel 6,000 in 1992, and supplier firms were either sold or closed down. A further reduction of almost a quarter in the number of jobs was anticipated, comparable to the shock on the labour market of the 1970s.

Lessons learned in the crisis of 1974 but, more particularly, new instruments developed with the trade unions during the 1980 crisis, which enabled companies to adjust employment peaceably by means of early retirements, not replacing 'natural wastage', a freeze on hiring or by reversing the decision to buy parts and components, may no longer be valid. In fact, a particular system of industrial relations had emerged in Germany in which workers were able to influence decisions on three levels, that is the works council on the plant level, the representation of workers in the boards of large companies (which is particularly true of VW) and trade unions on the regional and national level (Schregle, 1989; Streeck, 1984). This system has been known for its capacity for compromise and consensus, so that increasingly works councils and trade unions accepted some managerial responsibilities. There are, of course, important differences between large and small enterprises as well as between German (particularly VW which was trying to achieve the goal of employment stabilization for many years) and foreign ones, with USA companies facing larger changes in employment. In total, however, trade unions established instruments intending to achieve a rising labour productivity, such as early retirement rules and efforts to increase the qualification of the workforce. What has been characterized as the 'rationalisation pact' concluded between works councils, trade unions and management (Jürgens, Malsch and Dohse, 1989, 113), has contributed to a rejuvenation of the workforce and a larger proportion of highly qualified personnel as there has been selectivity in labour force reductions, discriminating against low skilled, female and foreign workers (Semlinger, 1990). This, in turn, fostered the 'socio-technical thinking' so particular on the German shop floor (Dankbaar, 1992). During the recent special boom after German unification, another outcome of this policy was that works councils preferred overtime and extra shifts rather than a limited expansion of the workforce.

In principle, the works council at the enterprise level of large multi-plant companies enables the labour force to inform and coordinate plant level works councils and to co-manage company policy. Increasing Europeanisation of companies, however, could not be resisted as the German system of co-management was not applicable. Tendencies towards stronger cooperation among European trade unions (Platzer, 1991) do not hold. The workforce, therefore, made attempts to counterbalance increasing Europeanisation of automobile companies with the Single Market in the early 1990s. For the first time in automobile production, VW works councils set up a European works council which was formally accepted by management in 1992. But it was soon used not only to strengthen the bargaining power of workers against management on the European company level but also to prevent erosion and relocation of functions from the parent plant to plants in other European countries. In accepting management plans to shift the entire production of the small Polo model from Germany to Pamplona in Spain the enterprise works council obtained a promise that production of the two 'core' models - the Golf and Vento - for the European market, as well as the competence for further design even of the Polo, will stay exclusively in Wolfsburg. These policies might have come to a standstill with the emerging crisis and increasing awareness of what it means to be the neighbour of low-wage countries in the east.

Things have changed fundamentally, not only with the emergence of the Single European Market, but more from the opening of eastern Europe, from east Germany unification, and with crisis. Current policies of works councils, therefore, are more oriented towards safeguarding a whole plant than to maintaining a given number of jobs. Works councils and trade unions face an increasing squeeze from different directions, and it was quite debatable whether the experiences and instruments of the past would be suitable for adjustment during the 1990s. Step by step, a large part of the regulations which had been won by trade unions in recent decades in West Germany but proved to be 'inflexible' in the current phase of structural change were altered. Agreements have been cancelled, leading to the recent introduction of team-based work in flexible production sites, the extension of shift patterns including Saturday as a regular working day, and new job enlargement. Works councils worrying about plant closures have agreed to these changes despite the resistance of trade unions (Semlinger, 1990). Many a 'first tier' components company establishing a satellite plant could enforce new regulations to achieve internal flexibility of labour more easily if management chose a rural location where trade unions' power is traditionally weak. Competition not only emerged between west and east German workforces, clearly shown by east German works councils agreeing more readily to innovations in the organisation of production than their west German counterparts. Imminent relocation of production to eastern

Europe and imminent further dismissals have resulted in agreements on completely new models of work time organization and reductions in the incidental wages costs which are the highest on the world. In consequence, companies achieved both a previously unattainable degree of work flexibility, of increasing productivity and reducing wage costs. BMW already had achieved agreements on work flexibilisation in 1991; other companies subsequently followed. As a result, in early 1994, after a site selection process including eastern and southern Europe, VW announced that it was to produce the new compact model of Audi at Ingolstadt whereas Mercedes-Benz allocated production of the new city car model to Rastatt and the new engine generation to its Untertürkheim plants.

These are indications that the consensus between labour force and management still exists and the core of automobile production in Germany will not be easily eroded in the near future. It is to be stated, however, that labour in other manufacturing sectors and in smaller companies does not have such opportunities to co-determine future processes of regional deindustrialisation and relocation. Consequently, the bulk of changes in the parts supplying industries may not be resisted strongly.

5.6 The Changing Geography of the German Automobile Production System During the 1990s

Changes of markets, of the locational pattern of automobile assemblers as well as of component suppliers and, finally, changes in the work organisation slowly, but increasingly, are determined by the integration of the German automobile production system into a larger European one. For as long as even the British and German plants of the USA company Ford procure 66 per cent and 70 per cent of components and materials nationally in the UK and Germany, Europeanisation will be delayed a little (Wells and Rawlinson, 1992). But further shifts towards European fringes result both from increasing pressures for competition from non-European companies as well as from concentration among European companies. Fig. 5.2 summarizes the increasing squeeze on the German automobile production system. Although this has already caused fundamental adjustment processes at German locations, different strategies emerging to safeguard the 'core' position of German locations within the evolving European system will call for further adjustment. One has been the recent takeover of Rover by BMW, another are takeovers and greenfield investments of German suppliers in western, eastern and southern Europe. Greenfield investments in east Germany which only started production in 1992 will remain the last ones for many years in Germany. In contrast, volume producers such as Ford and Opel, but increasingly VW, which had

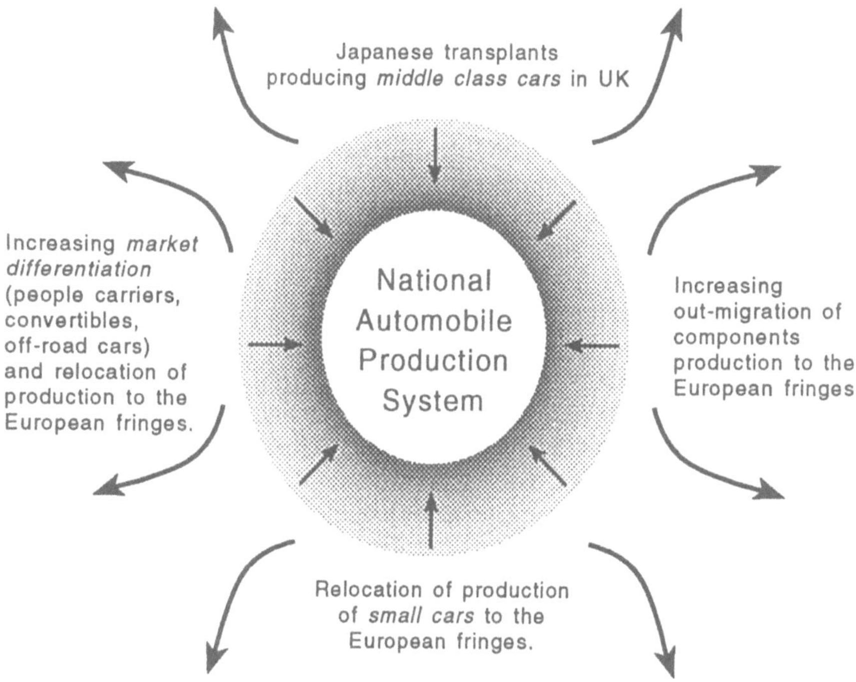

Fig. 5.2 The squeeze on the German automobile production system.

integrated foreign plants into a European production network during the 1980s will increasingly be followed by large component suppliers establishing production networks of their own. Hence, structural changes in the automobile industry and its suppliers stemming from both market penetration and differentiation strategies, as well as cost-reducing strategies of lean production and modular sourcing, will generate an increasing spatial division of labour on a European scale.

The consequences for the spatial organization of the automobile production system in Germany will be serious. Employment will be radically reduced with different implications both for all segments of the labour force and locations. Automobile assemblers will be less likely to close down plants but, in co-management with works councils and trade union, upgrade the plant's functions within the firm's own production network. Such is the case with all automobile assemblers enlarging R&D facilities and maintaining higher valued car production at traditional core sites. As a result, labour force qualification will matter even more in the future than in the past. Spatial agglomeration and specialization will probably grow in importance, as may be shown by the emergence of R&D agglomerations of suppliers and engineering firms at traditional production sites such as Frankfurt am Main, Stuttgart and Munich. On the other hand, rural supplier locations with poorly skilled jobs are seriously threatened by closure and shifts to east or west European countries. The spatial division of labour between German high-tech and assembly plants and medium and low-tech plants in other European countries will probably be deepened.

Finally, we have to bear in mind that Germany still will remain the largest regional automobile market in Europe. Clients may, however, increasingly shift their preferences from German makes. While west German clients still largely buy 'German', which means the models of VW, Opel, Ford, Mercedes-Benz and BMW wherever they may be produced, east German clients prefer cheaper cars, increasingly including Japanese and Korean makes. Thus, foreign automobile assemblers may strengthen market access, either in the design of new models via R&D, such as the Japanese firms Honda, Mazda and Mitsubishi running their European R&D centres around Frankfurt am Main or in sales via distribution subsidiaries which mainly cluster around the Rhine line from Cologne to Düsseldorf (where the German distribution subsidiaries of Renault, Citroen, Volvo, Rover, Chrysler, Daihatsu, Mazda, Nissan, Toyota are found). Changes in the spatial organisation of the automobile production system, therefore, will result in an increasing specialisation of urban agglomeration areas.

Chapter 6
The Italian Automobile Industry and the Case of Fiat : One Country, One Company, One Market ?

Sergio Conti and Aldo Enrietti
International Territories Department, Economics Department, University of Torino, Italy

6.1 Introduction

In the current period of globalisation of economic processes and the pervasive nature of information technology, both territoriality and national and/or regional comparative advantages have been simultaneously reinforced. They in fact assume a key strategic value in the "new" behaviour of actors performing on a virtually world stage. The analysis of the Italian automobile system (represented, in reality, by a single large company, Fiat) clearly exemplifies this tendency, which has come to prominence in debates in the social sciences: its competitive dynamic can, in fact, be interpreted on the basis of the relations which it has established, historically, with the national system (as well as local and regional ones).

The chapter begins by considering the composition and the fundamental structural dynamics of the industry, in the light of the inseparable relationship between enterprise strategies and national economic policies. After this, a synthetic reconstruction of the historical evolution of the spatial behaviour of the enterprise leads to a discussion of present processes (of rationalisation and of both national and international expansion) which see the Italian automobile industry involved in the search for new equilibrium.

6.2 The Italian Automobile Industry: Some Structural Features

The Fiat Group currently represents the largest Italian industrial group. A few figures will give an idea of its significance to the Italian economy: revenue represents 3.5 per cent of GDP, employees are 4.2 per cent of those in employment, investment is 2.1 per cent of the total, and research and development is 13.1 per cent. Fiat SpA is a holding company with diversified

Table 6.1 The Fiat Group in 1983 and 1991

Sectors	Returns (Billion Lire) 1983	1991	Employees 1983	1991	Investments (Billion Lire) 1983	1991
Car	11888	27506	116397	128925	902	2159
Industrial Vehicles	4517	8520	39013	41320	182	608
Earth Moving and Agricultural Machines	2539	2116	18304	10236	59	46
Metal Products	871	1193	14605	9888	36	88
Vehicle Components		3267		25997		239
Batteries		752		4875		32
Industrial Components		2184		9247		83
Components	2250		29052		125	
Production Means and Systems	465	1106	4366	5153	7	26
Civil Engineering	394	2307	3142	8567	22	45
Railway Systems	140	285	1331	1610	5	14
Aviation	342	1174	3485	4769	30	72
Telecommunications	340		4671		17	
Thermomechanic	332		1955		9	
Publishing and Media	193	404	1360	1250	16	8
Snia BPD		2318	111150			231
Intra-group		-4353	5925			172
Financial Services		2655	1896			36
Insurance		1748	2225			90
Main Distribution		4123	14874			234
Tourism	234		3022		24	
Holding and other sectors	188		2291		16	
TOTAL	21985	556550	243608	287957	1453	4183

activities (Table 6.1) but only fully adopted this structure at the end of the 1970s (Mosconi and Rullani, 1978). Before this, it was organised on functional lines, with a single profit centre and real decision-making power concentrated in the person of Vittorio Valletta. The transition to the holding company structure began in 1973 with the constitution of Fiat Engineering and ended in 1979 with that of Fiat Auto.

Despite diversification, however, company resources and skills are concentrated around the core business of the *vehicle industry* (Enrietti, Follis and Fornengo, 1988), to which the following Fiat sectors belong: automobiles, industrial vehicles, metallurgical products, vehicle components, industrial components, accumulators, means and systems of production. In 1991 these accounted for 82 per cent of overall revenue. The automobile sector alone generated 48 per cent. This bias towards the automobile industry deepened further during the 1980s through various acquisitions (Table 6.2).

The last decade has also seen diversification, in particular towards chemicals and the arms industry (Snia BPD), financial activities and large-scale retailing (Rinascente). Such diverse acquisitions nonetheless share two features in common: a focus on the Italian market and a not particularly high, though growing, turnover. With the process of growing internationalisation - which remains centred in Europe, however, where 90.6 per cent of employees, 93 per cent of production plant and almost 92 per cent of revenue are concentrated[1] - the Fiat Group as a whole has gradually reduced its dependence on Italy. Even so, in 1991 sales in Italy represented 53.8 per cent of the total and employees

Table 6.2 Selected acquisitions by Fiat during the 1980s

Sectors	Firms acquired	Incorporated into
Industrial Vehicles	Ford U.K. (1986) Pegaso, Spain (1991)	IVECO
Cars	Alfa Romeo (1986) FSM (Poland,1992)	Fiat Auto
Components	Solex and Jaeger (1986) from Matra (France), electrical parts and headlamps from Lucas (U.K.), Carello (Italy, 1988)	Magneti Marelli
	CEAC (France, 1990) and Sonnenschein (Germany, 1991)	Magneti Marelli
Earth Moving and Agricultural Machines	Ford New Holland (1990)	Fiat GEOTECH

[1]Fiat consolidated accounts, 1991. In the following pages, when not otherwise specified, the figures are from direct Fiat sources.

there were 60 per cent of Fiat's total (with 27 per cent located in Piedmont, 17 per cent elsewhere in the north and 16 per cent in the Mezzogiorno), though this was a sharp fall from 85 per cent at the beginning of the 1980s. This continuing dependence on Italy constitutes a weakness for Fiat in a scenario of market globalisation, an issue which will be examined below.

Fiat Auto is the most important company in the group and effectively holds a *monopoly position* in vehicle production in Italy (with the exception of a few bodywork specialists such as Pininfarina and Bertone, and a few producers of sports cars, such as Lamborghini). This dominance is the product of a history of continual acquisitions and takeovers. Without going back to the period between the two world wars, it is enough to note those of the last twenty years: Lancia in 1968, Ferrari in 1974, Alfa Romeo in 1986, Maserati and Innocenti in 1990[2].

This process has blocked all attempts by foreign manufacturers to set up their own plants in Italy: the acquisition of Alfa Romeo shut out Ford and that of Innocenti stopped potential Japanese competitors. The defence of its own national territory has also meant that, despite possessing subsidiary producers and licensed manufacturers in various continents, Italy remains the heart of Fiat's auto production system: in 1991, only 25 per cent of total production (638,000 cars) was manufactured by sister or licensed companies abroad, 244,000 of which were made in in Brazil.

There is a similar concentration of sales: 60 per cent of the cars sold in Europe were sold in Italy, while Europe absorbs 94 per cent of Fiat's exports. This dependence on the Italian market results from a strategic decision taken at the beginning of the 1980s when Fiat Auto was confronted with a series of critical problems: high debt and low capitalisation, delays in updating its model range, an over-extended international presence, specialisation in low horsepower cars, rigidity in industrial relations and significant losses of market share in Italy (Silva, Grillo and Prati, 1982; Comito, 1982; Becchi Collidà and Negrelli, 1986). The chosen strategy was thus that of a drastic concentration of its efforts in the European, and especially Italian, market[3]. In effect, starting from 1980, Italy has absorbed between 68 per cent and 72 per cent of sales in Europe (compared to 63 per cent in 1979), giving Fiat the opportunity to set prices and achieve a greater rate of profit than elsewhere in Europe. The dependence on the European market has at the same time reinforced the company's specialisation in the lower segments of the market, focused on smaller and cheaper cars, that is those most in demand in the domestic market[4].

[2] A 49 per cent share in Maserati SpA, and the purchase of 51 per cent of Innocenti Milano SpA.

[3] Fiat left the United States market, gave up production in Chile, Argentina and Uruguay, and sold its controlling share in Seat (Spain).

[4] In 1989 in Italy, the so-called A and B segments covered more than 50 per cent of the market, compared to 41 per cent in France, 26 per cent in Great Britain and 16 per cent in Germany.

This strategy proved successful in the 1980s when demand for cars grew more rapidly in Italy than in the rest of Europe (sales doubled in Italy between 1979 and 1990, from 1,191,000 to 2,344,000, while in Europe there was an increase of only 39 per cent). This allowed Fiat to increase its market share in Europe, to almost 15 per cent in 1989. There is an important qualification to this leading position, however, as Fiat models have high shares where there are barriers to Japanese cars: Fiat holds 21.1 per cent of their market, against Peugeot's 17.1 per cent and Volkswagen's 10.1 per cent. In contrast, its position is much weaker in countries where no barriers exist (only 4.8 per cent against Fiat's European average of 14.8 per cent (Mitsubishi Research Institute, 1990, 5)

In Italy, in particular, the direct import of Japanese cars is limited to 3,600 a year, although there is the possibility of importing a further 27,000 through other member countries of the European Community. In off-road vehicles, where there is no restriction since Fiat stopped producing them, Japanese producers have taken 70 per cent of the market[5]. Defence of the Italian market will, however, gradually diminish in future following the implementation of the agreement signed in 1991 between the European Community and Japan. This specifies that in Italy, at the end of the transition period (1993-1999), the share of vehicles imported directly from Japan can rise to 5.3 per cent, plus a further 5.5 per cent imported through European transplants. Over the last two years, moreover, Fiat Auto's capacity to penetrate the Italian market has fallen considerably, stabilising at around 45 per cent, with a clear loss of 25 per cent since 1986[6]. This has also had severe effects on the volume of production in the country: thus, since 1989, the year of peak production, there has been a drop in annual output of more than 300,000, with serious effects on costs and employment.

Compared to its competitors, Fiat is distinctive in possessing one specific feature, that of being able to depend on an important series of supply

[5]Despite the limitations, the Japanese car share has risen constantly, reaching 2.7 per cent of the market, with 62,117 vehicles sold in 1991, compared to the 8,662 vehicles and 0.5 per cent share in 1986.

[6]The net reduction in Italian vehicles' market share can be put down essentially to three factors:

a) an aggressive commercial policy on pricing and accessories by competitors, above all Ford;

b) the failure of sales of the Tipo, which should have been Fiat's best-selling model after the Uno and which instead fell from the 217,000 cars sold in 1988 to only 137,000 in 1991;

c) the poorer updating of models (the new Uno will come out almost two years late compared to plans and the Alfa 155, which came out a year and a half late), in contrast with the introduction of new models by competitors (e.g. Clio, Golf, Astra).

companies belonging to the Group itself. The company took a strategic decision over these activities in the late 1970s (Enrietti and Fornengo, 1989)[7]. It transformed several of its own plants into independent companies but also acquired other companies. A first phase of rationalisation was followed by a second in which greater importance was given to innovation and diversification in the automobile market. The resultant organisational set-up comprised a division into three sectors (metallurgy, vehicle components, industrial components), each led by a head company:

1 Teksid, which sold its metallurgical production to the state nationalised industry, and then specialised in the vehicle field (smelting of cast iron, aluminium and other alloys, pressing and working of steel). In 1989 it employed 11,070 people.
2 Magneti Marelli became, in 1987, a sub-holding company specialised in high technology systemic vehicle components. Its objective is now to increase its internationalisation to reduce its high dependence on Fiat (for example, via agreements with Matra in France and Lucas in the UK).
3 Gilardini, in contrast to the other two cases, is also involved outside the vehicle sector. It does, however, hold a very limited export share in this field (less than 5 per cent) and is highly dependent on Fiat.

6.3 The Relationship Between the Automobile Industry and Government Economic Policies

The automobile sector has been influenced both by specific industrial policies and by more general macro-economic policy, within a framework which leaves the market a free hand. Government has not played an active role in directing the sector, notwithstanding the fact that a small but important company like Alfa Romeo was publicly owned until 1986. Given the clear supremacy of Fiat up to 1986 and its later monopoly position, state intervention has often been created tailor-made for the company, resulting in a mixture of company strategy and government policy for the industry. Consequently, "any possible company

[7] In Europe in recent years, manufacturers have followed two trends: one of abandoning their presence in components, such as Renault, which sold Renix to Allied Signal, Sotexo to Epeda Faure, Ateliers de Saint Sturn to Fruehauf along with others, or British Leyland, which sold Unipart; a second trend was to increase and improve the presence in the components sector, and this is the case above all of Fiat, but also of PSA, with the creation of ECIA.

errors are covered by public intervention and are transformed into national errors" (Silva, Grillo and Prati, 1982, 37).

As regards industrial policy, there have been three fundamental lines of action: protectionism; hand-outs and bail-outs; and incentives intended to produce territorial re-equilibrium.

6.3.1 Protectionist policies

These have involved both export incentives, with the reimbursement of IGE (the predecessor of VAT) on vehicles exported and an additional tax on imported vehicles, varying between 4 per cent and 7 per cent. In addition, until 1958, cars and motorcycles were the most heavily protected items, with an average duty of 40 per cent, the highest in Europe, levied on imports. With the disappearance of duty within the EC, the agreement between Italy and Japan, reached before the formation of the EEC, limiting the annual import of cars to 2,200, remained in force. With the increase in imported cars, taxation was graded according to engine horsepower, with severe impacts on cars of over 2000 cc, most of which were foreign[8].

A further form of protectionism prevented direct foreign investment, so that in Italy foreign manufacturers have been unable to establish themselves. Since 1945 there have been very few agreements between Italian and foreign companies. In the 1960s, Alfa Romeo assembled a small car under license from Renault and, at the same time, Innocenti (which then produced motorbikes) also began production of small cars under license from British Leyland; this second agreement ended in the 1980s and the British company was replaced by Daihatsu for the supply of engines. Finally, in the late 1980s, Alfa Romeo signed an agreement with Nissan for the production in Italy of a modified version of the Japanese company's Cherry model: Alfa Romeo supplied the Alfa 33 mechanical parts and Nissan provided the bodywork. The operation was not a success because the style of the body was obsolete and alien to Italian taste while the revaluation of the yen increased the costs of parts from Japan.

6.3.2 Bail-out policies

In the 1970s the state-owned company GEPI, involved in the restructuring of private companies, intervened in Maserati and Innocenti-Leyland. Its most important operation, however, undoubtedly involved Alfa Romeo. Its survival

---

[8] In 1992, VAT for cars of over 2000cc was 38 per cent compared to an average VAT rate of only 19 per cent.

had been guaranteed by the state in the 1930s and its losses, which began in 1974, were subsequently covered by the state. It forms an emblematic case of the failed role of state-owned companies as an instrument of industrial policy. The state did not support Alfa Romeo investments made in the 1950s, as these were self-financed or funded with credit from German banks. Nor did it intervene in the 1960s to bring negotiations between Fiat and Alfa Romeo over their respective product specialisation to a successful conclusion. As a result Alfa Romeo decided to expand not only in sports and powerful cars, but also in the medium-low segments of the market - that is, in direct competition with Fiat - with the production of the Alfasud in its new southern factory of Pomigliano d'Arco (Naples) and later, at the beginning of the 1980s, reached a joint agreement with Nissan for a joint car production venture in Italy. These two competitive "successes" of state industry over Fiat were not, however, the product of an overall policy for the auto industry. The first was the result of political and trade union consensus for the development of the Mezzogiorno while the second arose from Alfa Romeo's need to survive. The fact that state intervention has not taken a coherent form is also demonstrated by law 675 of 1977 on "Industrial Restructuring and Reconstruction". Although the difficulties of the automobile industry were evident in 1977, it was not included among the sectors to be restructured. It was not until 1980, when the Fiat crisis became acute, that it was decided to prepare an Automobile Plan and to extend the provisions of the law in question to this sector. Despite the delay, 32 per cent of the funds distributed by the state were allocated to the automobile industry by the end of 1984 (Momigliano, 1986, 44). The industry financed 57 per cent of its investment up to 1985 through law 675.

6.3.3 Policies of territorial re-equilibrium

Until the end of the 1960s, the automobile sector was totally concentrated in the northern regions, in particular with Fiat in Piedmont and Alfa Romeo in Lombardy. The first significant change came in 1967 with Alfa Romeo's decision to produce its new model in Naples, a decision linked to the fact that a law of 1957 constrained state-held companies to locate 60 per cent of their new plant in the south. A little later, Fiat began a policy of expansion in the south, under the social pressure of tension in the labour market and an assessment that the social costs of concentration[9] were being transformed into internal company costs. It was also stimulated to make this decision by new legislation on incentives. Law 717 of 1965 extended benefits previously limited to small

[9]The population of Turin rose from 1.4 million in 1953 to 2.3 million in 1971, with a significant delay in the supply of housing and public services.

and medium size enterprise to large-scale projects. Thus, in just four years, eight new Fiat southern factories were created: Termini Imerese and Cassino (car bodywork), Termoli and Sulmona (car engineering), San Salvo (components), Bari (fork-lift trucks), Lecce (earth-moving machinery), and Brindisi (aviation). At the end of the 1970s, further investment in new production capacity was made by Fiat in the south: Foggia (diesel engines for industrial vehicles and cars), Grottaminarda (buses), Atessa (commercial vehicles, in a joint venture with Peugeot).

The next substantial round of investment in the south by Fiat Auto came in the early 1990s; in particular[10], two new factories in Melfi (producing the Punto, the replacement for the Uno) and Avellino (components). This is an overall investment of about 6,300 billion lire, of which the state will contribute 1,470 billion in grants over five years and a further 1,630 billion to cover interest repayments. More generally, the Fiat group has also planned to move part of its research activities to the Mezzogiorno, with the creation of new laboratories, one specialised in research in the car sector in Pomigliano d'Arco and another for the design of expert systems for cars, a software centre for traffic control in Salerno, a testing centre for road vehicles in Nardï (Lecce), and, finally, a research laboratory for civil aircraft engine components in Brindisi.

6.3.4 Other forms of state intervention

State intervention of a general nature has had significant effects on the automobile industry in general and on Fiat in particular. Firstly, the expansion of the motorway network, which grew from 1,169 km in 1960 to 5,760 km in 1978, making it one of the largest in Europe, was an indirect stimulus to road transport. It was accompanied by a clear policy of favouring private over public transport, reducing investment aimed at improving the efficiency of the railway system. Another instrument of industrial policy, law 46 of 1982, which instituted a Fund for Technological Innovation, was widely used by the automobile industry: 34.8 per cent of all the programmes financed up to 1985 was allocated to the automobile and components industry (Momigliano, 1986, 156). Finally, the Cassa Integrazione Guadagni (state lay-off and unemployment benefits based on previous wage levels), an instrument used mainly to make redundancy socially acceptable, played a significant role in shaping the course of the restructuring of the 1980s. Again in this case, the automobile industry enjoyed disproportionately greater benefits than would

[10] It should not be forgotten that, with the purchase of Alfa Romeo, Fiat Auto had also obtained the Pomigliano d'Arco plant in Naples.

have been expected in relation to its share of industrial employment (Saraceno, 1986, 77-8).

This section has provided an overview of the Italian automobile system and of the Fiat group in particular, exploring the complex dialectic between state economic policies and company strategies. This enabled the identification of certain important specific features (and anomalies) of the Italian "model" of automobile production. In the next section, a rather different perspective will be adopted.

6.4. Spatial Strategies and Reorganisation Strategies

6.4.1 The approach: the automobile industry as a complex industrial system

We shall now look directly at aspects of the territorial organisation of the company, going back to the fundamental bond that exists between strategic behaviour, industrial policies and the spatial dimension of development. The result that will emerge is a model of the production system within which it will be possible to reorganise meta-theoretically the wide variety of questions discussed in the two previous sections and to open up new discussion. The territorial approach to the problems of transformation of an industrial system seeks an holistic understanding of how the system as a whole "works", through the relations which link economic, social and institutional actors.

The focus will therefore be the evolution of the strategic behaviour of the Italian automobile sector, seen as a *system*; that is, as a whole made up of elements (units), of functional relations between them, and of interactions between the elements and the environment. In this light, the *units* can be appropriately analysed as a chain of activity through which the interacting subsystems contribute together to the generation of product value. Analysis of the automobile *filière* - "an organisation of companies based on the division of tasks and on bonds of co-operation and dependence" (Fenneteau, 1985) - is consistent with this objective and has the advantage - adapting Porter's (1990) scheme - that it can be extended to the analysis of inter-company functional relations. Irrespective of whether the latter are technical or organisational, they define, as they develop, a specific geographical configuration of the system's activities.

This type of relations cannot, in holistic approaches, be discussed independently of the relations between company and environment. In turn, these can be be divided into two main subsystems: the *competitive environment*, in which both the overall dynamic and the competitive behaviour of the actors within it take place; and the *economic and social environment* (local, regional) in which they operate. These now-consolidated concepts (units, relations, environment) will be instrumental in throwing light on the contemporary strategic behaviour of the Italian automobile industry. This is, in fact, an historical phase in which business behaviour is structurally framed within a dynamic of relations between global and local processes of a non-reducible complexity. It cannot therefore be interpreted through the all-encompassing (and formal) schemes which had previously dominated theoretical thought and empirical research.

The contemporary competitive environment, in which companies are increasingly organised in flexible, changeable structures, and in which technology and the environment itself become a generalised force of production, does not negate the previous "model" of the company as an hierarchical organisation coming to grips with oligopolistic competition and the guiding and co-ordinating role played by the state. The contemporary industrial "model" becomes dominant, but does not obliterate the previously dominant organisational "models". It is for this reason that periodisation becomes a decisive theoretical tool for the study of the system's organisation trends: they are, in fact, dependent on the forms of expansion, change and bifurcation which have preceded it.

6.4.2 From expansion to crisis

The years immediately following the second World War were crucial for the future development of Fiat. Within a rapidly expanding national market, without any significant import penetration and with few, and modest, local competitors, Fiat focused its market strategy on small and medium-sized models. It also organised production in a way which paid little attention to R&D, planning or scientific activity. Fiat's entire strategy was directed towards *quantitative expansion*, both in terms of employment and productive capacity [11], paying little attention to its internal organisation.

These technological and product decisions were accompanied by a spatial strategy which deliberately enhanced the relative advantages of concentration and mass production. This had two main features: concentration in the north-west of the country and in the city of Turin in particular; and concentration of production in a few large, vertically integrated, factories. In the Turin area, Fiat has in fact found, produced or has seen others (the public administration)

---

[11] Between 1952 and 1960, the number of cars produced in Italy by Fiat rose from 101,660 to 515,315, reaching 1,450,600 in 1968. Employment grew in the same period from 71,859 (1952) to 92,891 (1960) and 158,445 (1968).

produce a large part of those "territorial conditions" which marked the establishment of the system of mass production (Castronovo, 1971; Gabetti, 1977). Turin is rightly seen as representing the model of the factory town, albeit with its own specific features and connotations. It comes as no surprise that even in the international literature Turin has been a favoured area of study for the relationship between industry and the city in the era of the second industrial revolution (Gabert, 1964; Jalabert and Gregoris, 1987; Sallez and Schlegel, 1963).

This technical and location choice defined a company strategy which was not to change until the late 1960s. With a few exceptions and timid processes of internationalisation, Fiat's industrial investment was identified with the boundaries of the Turin agglomeration: this area not only contained the entire car production cycle (in 1968, about 1,300,000 out of 1,550,000 cars came from the Turin area), but Fiat also participated in the whole range of production based on the internal combustion engine, involving a close-knit network of small and very small supply companies, often completely dependent upon Fiat. While at the end of the 1960s, there were over 125,000 employees working in Fiat's Turin factories, at least an equal number worked in production units which directly or indirectly were part of its network of sub-contractors. In reality, it has never been possible to delimit exactly the boundaries of this network because its composition is in a state of constant flux, especially as regards the smaller suppliers. Fiat drew on around 1,200 direct contractors, about a third of the 3,500 units (often small and very small) linked in some way to the automobile industry (Rossignolo, 1971). Under these conditions, despite its considerable internal vertical integration, during the 1960s Fiat acquired on average over 50 per cent of the value of its total turnover from external companies, of which about half were located in the Turin agglomeration. As a consequence of the tight dependency relationship between Fiat and its myriad of suppliers in the area - a relationship which often included even financial control - the concentration of capital and property in Turin came to exceed even the degree of sectoral specialisation [12].

As a result of the automobile industry's strategy towards indefinite output expansion and a consequent, and sometimes uncontrolled, enlargement of the productive capacity [13], the Turin area became one of the most sectorally

[12] At the end of the 1960s, the spatial concentration of Fiat's activities in Turin was almost absolute. The few decentralised plants were mainly non-car producing, e.g. agricultural tractors or industrial vehicles, and in most cases they were the result of takeovers of pre-existing firms rather than Fiat's locational decisions. In 1968, Fiat's presence in the south of Italy was restricted to two small production plants in Naples (700 employees in all), the retailing outlets and customer service centres and shareholdings in a few firms (Spriano, 1985).

specialised regions in Europe, comparable to just a few other international examples such as Detroit and the West Midlands. In Turin, in contrast to these other areas, all these activities were concentrated within a single corporation, while the company's control extended deeply into the labour market and was at the centre of intense interlinking of political and economic roles. It in fact strengthened itself through the regulation of financial processes, the mechanisms of successive sub-contracting relationships, a monopoly over research, managerial assistance largely internalised within the one large corporation, and through privileged relationships with both the local and obviously national political systems (Borlenghi and Dematteis, 1982; Conti, 1986).

6.4.3 The first turning point

The first great wave of reorganisation of the Italian automobile industry - and of Fiat in particular - started in the 1970s and can be largely explained by the *environmental* crisis produced by the locational and organisational "model" followed until that time. The rigidity of the connections between the city and the company - along with a more generalised crisis in international demand, especially in the market segments in which Fiat specialised - created conditions which prejudiced the network of co-operation which had previously ensured successful accumulation. In that period, not only Turin, but also other major auto production centres such as Wolfsburg, Lordstown and Boulogne-Billancourt were showing the same symptoms. However, the intensity and the duration of the crisis in the Italian factory-town hinted at a particularly serious state of social congestion. This produced a series of negative effects, so that the environmental diseconomies became the cause of instability and crisis for the company. It developed, in fact, an entropy of the social environment which made the process of change triggered by the company increasingly difficult to foresee and control.

For Fiat, this form of production rapidly became impracticable, not because of an ageing technological base but principally for reasons of a social and political nature. The reorganisation of the sector thus implied a profound change in the economic and territorial development "model", expressed in:

---

[13]In 1969, Mr Giovanni Agnelli - Fiat's chairman - forecast an expansion of the domestic car demand of 500,000 units in the following ten years, and that in 1978, taking into account the import/export dynamics, the market would have absorbed more than two million cars of national production. In reality, this number came to be less than 1,500,000 mainly due to the less rapid growth of the domestic market and to higher import penetration.

- a profound internal reorganisation of the conglomerate, involving the transformation of Fiat from a traditionally integrated company, with a very rigid and hierarchical managerial structure where conflicts of responsibility and interest often arose, into a "divisional" structure. Initially, in 1972, it was split into three operational sectors (cars; industrial vehicles and tractors; miscellaneous activities) and then in 1976 into eleven sectors, each headed by a holding company, in turn controlled by a single central holding company (the situation in 1991 is summarised in Table 6.1);
- in constant attempts to regain control of the production process through the introduction, in the "old" Turin factories, of electronic technologies, accompanied initially by considerable falls in employment;
- in a changed policy of industrial location, put into practice in 1970 with a two-year investment plan for the construction of nine factories in southern Italy, with about 17,000 employees. Employment in them doubled by 1981, mainly financed by regional policy legislation in favour of the Mezzogiorno. The other element of the new locational policy involved even more substantial projects on an international level. During this period construction was completed of a car factory in the Soviet Union, and a share of Citroen was purchased. In 1971 an agreement was signed to build a new production plant in Poland, and a massive plan for investment in Latin America (mainly Brazil and Argentina) began to be implemented. In western Europe, most production remained in Italy and Spain (Seat), but with significant exceptions in Ireland and Portugal. This intense process of internationalisation meant that production of Fiat vehicles abroad (in production and assembly plants, in subsidiary and licensee companies) reached high levels: 47.9 per cent of total production in 1980, against 17.2 per cent in 1968. The origins of Fiat's international expansion strategy lay in acceptance of the then-dominant notion that the only survivors would be the manufacturers producing around two million vehicles a year, the level seen as necessary in terms of economic, financial and technical criteria. Given the size of the domestic market and the limited scope for export growth, international expansion of production, above all in developing markets, was seen as the only path to follow. Nonetheless, Fiat continued to depend heavily upon the domestic market, where its competitiveness and profitability were higher.

The overall goal of reorganisation was to seek maximum flexibility in production methods and locations, which in turn implied operational objectives: first, to move into new market areas and, later, to structure the decentralised plants *as a function of the whole system*. The new organisation of production was thus not limited to the mere duplication of existing plants, but was aimed at decentralisating specific stages of production to dispersed medium-sized and specialised factories which were functionally and strategically connected (Amin, 1986). The strategy was, in other words, to "lighten" the Turin area - accompanied by the standardisation of some manufacturing processes and some intermediate products without undermining the continuing technological and organisational "centrality" of Turin.

The decade was, at the same time, one of uncertain and contradictory strategic choices by the Group as a whole (Comito, 1982, 212). Initially (as with the programmes in the Mezzogiorno, the acquisition of Lancia and international investment), the company still gambled on ambitious quantitative expansion[14]. In contrast, in the second half of the decade, just when Volkswagen and Renault were planning and producing ambitious relaunch programmes for the product segments in which Fiat held its largest market share, the Italian company announced various diversification projects which appeared partially to question its historical role as an automobile producer. While there was talk of consolidating investment in the auto sector, priority was given to "non-automobile production" (industrial vehicles, machine tools, bio-engineering, metallurgical components, earth moving machines and telecommunications).

This strategy was grounded in an interpretation of the effects of the first oil crisis which viewed cars as a "mature" product, the production of which was thus destined to be transferred to developing countries. Existing activities in industrialised countries such as Italy would have to be converted. Two product areas were favoured in the vehicle industry. The first was industrial vehicles, with the formation in 1975 of Iveco through the acquisition of the German company Magirus. The second was components, the importance of which increased following the construction of a steel works alongside the car plant in Brazil and the acquisition of Italian companies, in particular Gilardini. There was also further differentiation into "new" sectors, such as services and infrastructure. The consequence was that, for the whole decade, car production declined constantly, from 1,633,100 vehicles in 1970 to 1,278,000 of 1980[15]. Fiat's market share in Italy fell from 74.9 per cent (1968) to 51.5 per cent (1981) and some countries and markets were abandoned completely. In 1980, the agreement with Seat was broken, the attempt to enter the USA market was

---

[14]This wave of rationalisation investment was carried out, in fact, in years of great success for the company which, in 1968, had placed itself as second only to Volkswagen in world car exports. Again together with Volkswagen, it was the leading manufacturer in the EEC and controlled a market share of about 18 per cent, although it should be underlined that the size of these cars was much lower than the European average, and its share expressed in terms of sales was therefore less than 18 per cent of the total (Comito, 1982, 204).

[15]These figures have only relative significance, considering that the whole of the world automobile industry had to cope with unstable, segmented and less expansive demand from the mid seventies (on average annual growth rates of 8.5 per cent of the world market in the sixties were followed by figures of 2.7 per cent in the seventies, while demand actually fell for four years after 1979).

abandoned and an agreement was signed with Peugeot which foreshadowed withdrawal from production in Latin America.

6.4.4 The 1980s: the great rationalisation

At the beginning of the 1980s, then, Fiat Auto found itself in difficulties in its national market. It was losing market share and domestic demand for cars was growing more slowly than in other European countries (only in 1979 did demand in Italy return to the pre-crisis sales levels of 1973, while in the rest of Europe this happened in 1976). Further problems arose because of low productivity and difficulty in managing labour, the inadequacy of the components supply industry, absence of an industrial policy, and high inflation. In order to regain operating conditions comparable to its competitors, a complex defensive strategy aimed at restructuring was designed (Balliano, 1986), the main elements of which were:

- a cost-cutting policy: the first primary feature of this was to cut labour costs: 23,000 workers were made redundant and in 1980 the company started reducing its workforce by over 40 per cent in the next seven years, from 134,621 in 1980 to 77,910 in 1986. As a result, productivity levels nearly doubled, from 14.8 cars per worker in 1979 to 27.9 in 1985. By 1986 the relative share of labour costs had dropped to 18 per cent (compared to 27 per cent in 1979). The geographical distribution of employment shifted southwards, with the main losses in Piedmont, whose share fell from 78.8 per cent in 1979 to 72.6 per cent in 1985, while in the South employment increased from 16 per cent to 20.3 per cent.
- a technological strategy: technological transition was ensured by an intense process of fixed capital investment. It was maintained throughout the 1980s, sustained in part by three successive recapitalisations and by the sale of shares in the Spanish Seat company. The number of robots in use rose from 225 in 1980 to 2,500 in 1992 (and will rise to over 3,000 with the new factories in Melfi and Avellino); in some factories and some production segments automation was almost total (such as in the sheet-metal working stage in Cassino, south of Rome). A more detailed summary of the automation process is given in Table 6.3).

  The sudden introduction of such new technology necessarily had to be accompanied by plant reorganisation. This led to the almost immediate closure of three engineering factories, including the "historical" plant of Lingotto in the heart of Turin, the architectural structure of which did not allow the introduction of automated processes. On the whole, this reorganisation of the production structure led to a reduction in the number of active factories, to an increase in capacity utilization in those remaining and, above all, to the reduction in the break-even production point from the more than 1.5 million cars of the early 1980s to the 1.2 million at the

Table 6.3 The process of automation in Fiat Auto

1972	welding robot for the 132
1973	welding robot for the 131
1976	Digitron system for the automatic start-up for 131 bodywork; robot for underbody anti-vibration paint; installation of Robogate in Rivalta for automatic body assembly
1977	robot for first coat painting
1978	robot for enamel painting; Robogate welding system for the Ritmo
1979	Robogate welding system for the Panda
1980	LAM (asynchronous engine working) for engine frame assembly of the Ritmo
1982	Robogate system for the Uno; robotised welding of the rear axle of the Uno; robotised assembly of the cylinder heads for the Ritmo; robotised assembly of the complete steering column for the Uno
1984	robotised working and assembly of the Fire engine frame and head
1985	flexible robotised lines for the sheet metal assembly of the Therma and Croma; robotised Digitron system for the coupling of mechanical parts on Therma and Croma bodies
1986	flexible lines for complete assembly of mechanical gear parts; robotised welding and assembly of rear axle
1987	automated final assembly of the Tipo
1988	start-up of an automatic line, with 7 transfer presses, plus an FLM (Flexible Manufacturing System) for the production of cylinder heads for the Croma diesel engine and intake manifolds for the Therma and Croma
1989	further development of the rationalisation programme for the body shops at Mirafiori, Arese and Pomigliano d'Arco
1990	new paint and sheet-metal working shops at Pomogliano
1991	prototype of innovative presses for moulding of plastic materials at Mirafiori
1993	start-up of the "integrated" factory at Melfi

end of the decade (Scott, 1991, 258). In addition, the decision to allocate highly innovative investments to the South (automated and robotised manufacture and assembly in Termoli and automated vehicle assembly in Cassino) rather than Piedmont, as well as the closure of two plants in the North in 1992, at Desio (Milan) and Chivasso (Turin), has further reinforced the move of the centre of production towards the South. The overall strategy of decentralisation to the South can be attributed to four factors, in part already mentioned: the desire to reduce the degree of vertical integration and to decentralise so as to lower the growing diseconomies and problems of governability in the metropolitan plants; a redefinition of industrial policy towards the Mezzogiorno, more open to providing incentives to large companies; the reduction in the gap in costs (external diseconomies) and in efficiency between the North and South, a reduction on which the Fiat plants themselves had a further effect; the availability of both uncongested areas and labour. However, the central company functions (top management, research, management training, purchasing) have remained concentrated in the North; it was only in the 1990s that some research functions began to be moved to the South.

- The restructuring of supply: a further problem was the generalised reorganisation of the supply system, given that Fiat currently has a level of vertical integration of about 45 per cent of turnover. It purchased 13 per cent of requirements directly abroad, another 25 per cent from Italian branches of foreign multinationals, 35 per cent from by independent Italian manufacturers and 27 per cent from elsewhere within the Fiat group. As we have seen, this system was broken down into a large number of small-scale units, specialised in the production of relatively simple components with low technological content: it was, in other words, a system largely incompatible with the rapid transformation of models and technological innovation.

From 1980, Fiat began a drastic selection process on the lines of the Japanese "model" based on the ability to ensure innovation, competitive prices, and reliability. It resulted in the disappearance between 1980 and 1982 of about 350 direct suppliers, the number of which fell from approximately 1,200 to 850 companies (Enrietti, 1987). More particularly, the restructuring of the supplier chain was achieved in a number of ways. First, the instigation of a three-level hierarchy of component suppliers, each with distinctive roles: producers specialised in complex components (macro-component systems), which possessed a considerable capacity for product development, for independent generation of research programmes and for international competitiveness. Close collaboration was then developed with these suppliers in product design and development. The second level comprised suppliers of simple components, almost exclusively for automobile use, which in turn supplied the specialised producers. Finally, there were suppliers of standard and multi-purpose products. Secondly, by containing price rises below the rate of inflation. Thirdly, via the launch of a campaign of "development contracts", on the basis of which Fiat contributed part of the investment for the development of new products, guaranteeing itself the property rights to the results of the design and engineering activities. Fourthly, by introducing long-term supplier contracts, lasting from 3 to 5 years, for companies whose "development contracts" had been a success and for technological leaders. A further element which characterised the process of restructuring of relations with suppliers concerns the organisation of supply according to the principles of just-in-time. While initially emphasis was placed upon the maintenance of suppliers' warehouses, the most recent trend is towards synchronous JIT production, with the disappearance of supplier and manufacturer stocks. As a result, component suppliers have to locate their plants close to the assembly plants. The Melfi plant, in which the Punto, the car replacing the Uno will be assembled, is a case in point: alongside the Fiat plant there will be an industrial estate of about 600,000 sq.m. Sixteen suppliers, most closely connected to Fiat, will locate there. The components that they will produce are press products, both metal (exhaust systems, sub-groups, suspensions and brakes, window devices, brake levers, hinges) and plastic products (dashboards, bumpers, tanks, fittings, roof and door panels, seats). In some cases (roof panels, silencers, dashboards, seats), because of the high volume and low value of the components, the solution adopted will eliminate high transport costs. Likewise, the area of the Chivasso Fiat factory, which is now available after the plant closure, is going to be passed on to Piedmontese Fiat suppliers.

Further developments seem to suggest a greater spread of multi-plant enterprises, with geographically separate planning and production, as well as extending the just-in-time approach to the suppliers of Fiat Auto's suppliers. In general terms, the trends of the 1990s for component supply can be summarised as follows: a move to the south of components production plants, following the new Fiat plants; greater internationalisation of Fiat purchasing from its current low level (about 13 per cent), although the recent devaluation of the lira will slow down the process; a further selective reduction in the number of suppliers: by 1994, Fiat Auto expects that there will be no more than 500, roughly one supplier for each product line, of which there are 440. This means a closer co-operative relationship and partnership between the two parties, giving suppliers wider planning functions.

6.5 Continuity and Discontinuity in the Geography of the Italian Automobile System

With the gradual rise in the number of production plants in the South (see Table 6.4 and Figure 6.1), this area has become increasingly important in the dynamics of the technological and organisational development of Fiat Auto. During the 1980s, with the factories in Termoli (for the Fire engine) and Cassino (for the production of the Ritmo and later the Tipo), the Highly

Table 6.4 The location of Fiat plants in Italy

	1971		1980		1992		1994	
	n°	per cent	n°	per cent	n°	per cent	n°	per cent
Piedmont	6	47	7	46	3	20	3	18
Other Northern Regions	2	15	2	14	1	7	1	6
Central Italy	3	23	3	20	4	26	4	23
South	2	15	3	20	7	47	9	53
Total	13	100	15	100	15	100	17	100

Source: Fiat data

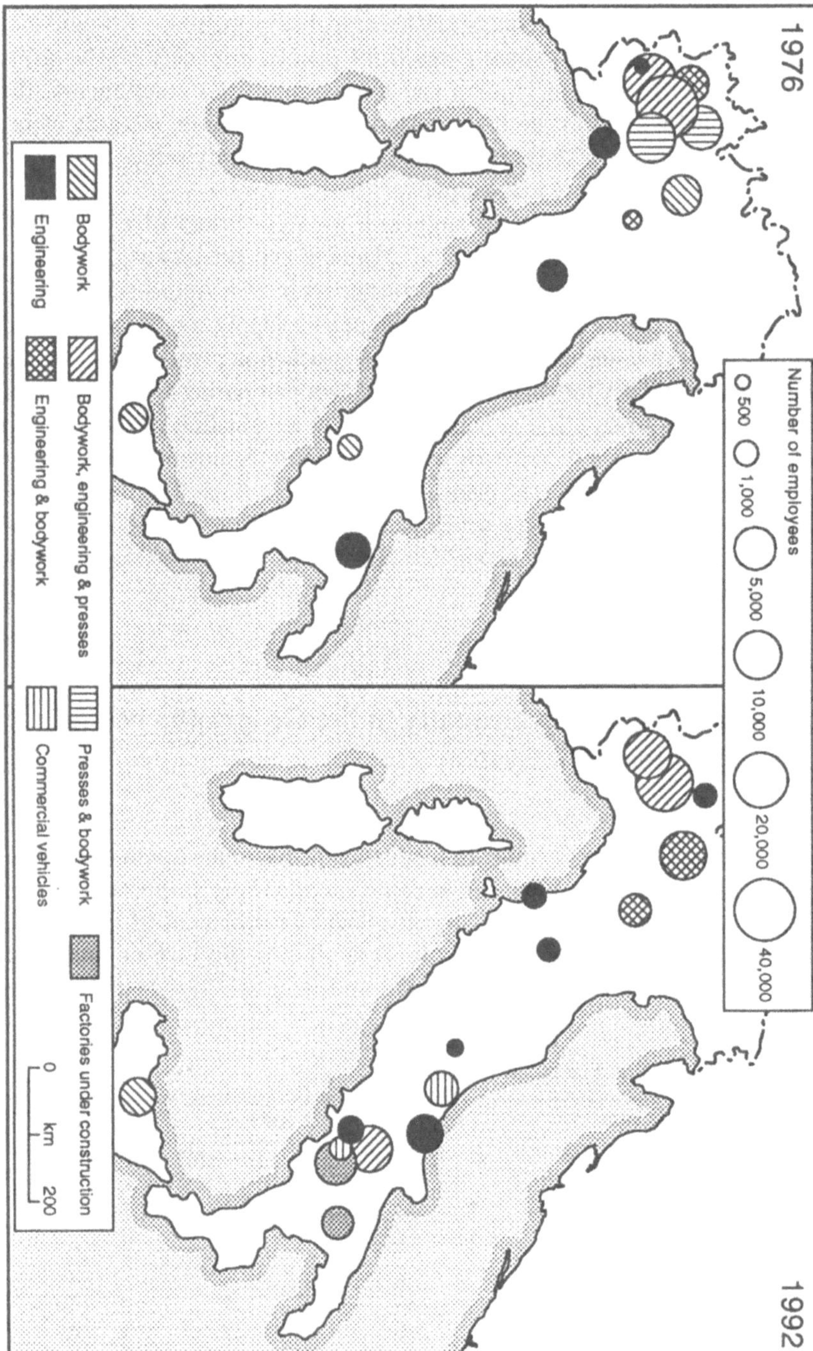

Fig. 6.1 Location of the automobile industry in Italy, 1976 and 1992.

Automated Factory (HAF) was introduced. This marked the passage from the traditionally rigid automation to a flexible form. This strategy corresponded to Fiat's hypertechnological decision that new technologies could solve the problems of rigidity in the Fordist model.

The reorganisation of the Cassino and Termoli plants was in fact directed at obtaining high flexibility, allowing the possibility of producing more models and versions in the same factory at the same time, increasing the degree of differentiation of the various models, and adapting them to the needs of specific market segments. Thus, at Cassino, flexibility, in terms of the possibility of alternating production of different models during the same day, at the end of the 1980s was almost double that of the "old" Rivalta plant, in the Turin metropolitan area, where the same car models were produced. Unsurprisingly, productivity analyses carried out in 1988 on a sample of 38 world car production plants (Krafcik, 1988) showed that Fiat had the best performance in Europe, even though there was still a gap between it and Japanese manufacturers.

By the end of the 1980s, however, the HAF model began to be questioned, as it was realised that Total Quality, Fiat Auto's strategic objective for the nineties, "cannot be the result simply of high technology, but also [must be] the fruit of intelligent and responsible human work" (Bonazzi, 1993, 77). Thus Total Quality was incompatible with administrative Taylorism, a powerful hierarchy and a marked subdivision in duties, all still present in the HAF. A new technological and organisational strategy was thus devised. This was the Integrated Factory (Cerutti and Reiser, 1991; Bonazzi, 1993), in which the

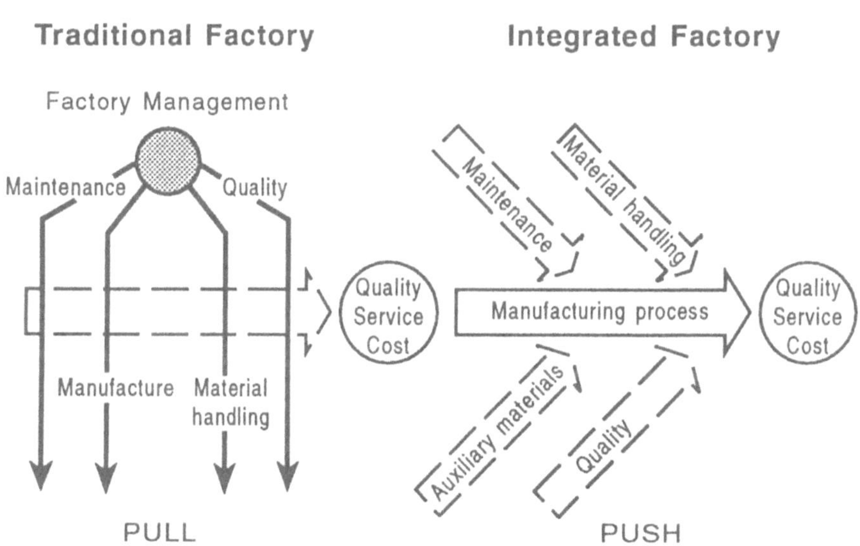

Fig. 6.2 Towards the integrated factory. (Source: Fiat Auto).

adjective integrated underlines the project's main feature and aim: to integrate the functions present in the factory by process (for a brief illustration of the Integrated Factory, see Figure 6.2). Drawn up at the end of the 1980s, the IF project was first introduced experimentally in the plants at Termoli and Cassino, in which the HAF had already been introduced, because this was an organisational model closer to the IF than the traditional northern Italian plants. But it will be above all the two new factories in the South (Melfi and Avellino) that will most fully represent the IF model, because they were designed according to its principles. Moreover, they are greenfield plants, for which the workforces have been very carefully selected, specifically to meet the requirements of the IF. All 8,300 employees will follow training courses; about 10 per cent of the office workers will be graduates, reflecting the high level of professional qualifications demanded by the IF; office workers will work in the plants, alongside production lines, thus eliminating the traditional office blocks. Unlike those built in the 1960s and 1970s, these factories cannot be considered additional to but as a replacement for employment in the North. This will impact on component suppliers, many of which, as we have seen, need to be in the proximity of the assembly plants[16].

Of particular importance in this context is the logic which has driven Fiat to locate those new plants which will become operational between 1993 and 1994 in southern Italy. Initially other countries were considered, particularly those in the Mediterranean area (such as Spain, Turkey, Portugal and Greece), which offered investment incentives. These were analysed on the basis of a series of factors: the country-system, financial incentives, infrastructure, the presence of component manufacturers and labour costs (lower elsewhere than in Italy)[17]. The ultimate choice of Southern Italy reflects the fact that, for Fiat Auto, this region was already an integrated area, both in terms of final production activities and research centres and a consolidated local network of component manufacturing companies. Also of considerable importance was the possibility of employing young people and workers with a high school leaving certificate, thus avoiding a working class with a long and consolidated tradition. A further attraction was a union agreement which allows a three shift working for six days a week.

The overall rationalisation of the system would thus seem to be going in the direction of a production structure based on new forms of "polarisation" around a few highly integrated plants: Mirafiori and Rivalta in the Turin area, Arese in

[16]The most recent figures available on the geographical distribution of purchases in Italy refer to 1985 and show Piedmont at the top of the table with 54.4 per cent and Turin alone with 46.3 per cent followed by Lombardy (20 per cent). The southern regions had only the very modest share of 9.5 per cent.

[17]For example, if labour costs are rated as 100 in Italy in 1990, in the Mezzogiorno they were 83.2, 25.7 in Portugal, 52.5 in Greece and 69.1 in Spain. Source: Svimez, 1992, 115.

Table 6.5 Geographical distribution of the main production phases in Italy

	1971			1980			1992			1994		
	n°	per cent		n°	per cent		n°	per cent		n°	per cent	
Body Work												
North	6		60	6		60	3		33	3		30
Centre	2		20	2		20	2		22	2		20
South	2		20	2		20	4		45	5		50
Total	10	45	100	10	44	100	9	41	100	10	40	100
Engineering												
North	5		63	6		67	4		40	4		36
Centre	2		25	2		22	3		30	3		28
South	1		12	1		11	3		30	4		36
Total	8	37	100	9	39	100	10	45	100	11	44	100
Presses												
North	4		100	4		100	2		67	2		50
Centre	-		-	-		-	-		-	-		-
South	-		-	-		-	-		-	-		-
Total	4	18	100	4	17	100	3	14	100	4	16	100
Total Italy	22	100		23	100		22	100		25	100	

Lombardy (formerly Alfa Romeo), which, with the southern factories of Cassino, Termoli, Melfi, Pratola Serra and Pomigliano d'Arco (again formerly Alfa Romeo), will constitute the key nodes in the Italian automobile production system. The expected effect of this, therefore, is the creation of a strategy of systemic integration between the various Italian regions and within the Mezzogiorno itself by creating greater uniformity in the distribution of the phases of work throughout the territory. In conclusion, more detailed analysis of the production stages in each plant[18] (Table 6.5) lends support to these claims: as far as the stages of bodywork and sheet steel pressing are concerned, there has, over time, been a process of replacement of the plants located in the North with ones in the South. Engineering is the only activity in which the number of plants has actually increased with the new factories in the South. It follows that in 1994 there will be an equal number of engineering plants in the two parts of the country, while there will be more bodywork plants in the South.

[18] Some plants have a single specialisation, such as Cassino (bodywork) and Termoli (engineering), while others work in several production stages, such as Mirafiori in Turin, Rivalta and Pomigliano d'Arco (engineering, bodywork and presses) and Arese (bodywork and presses).

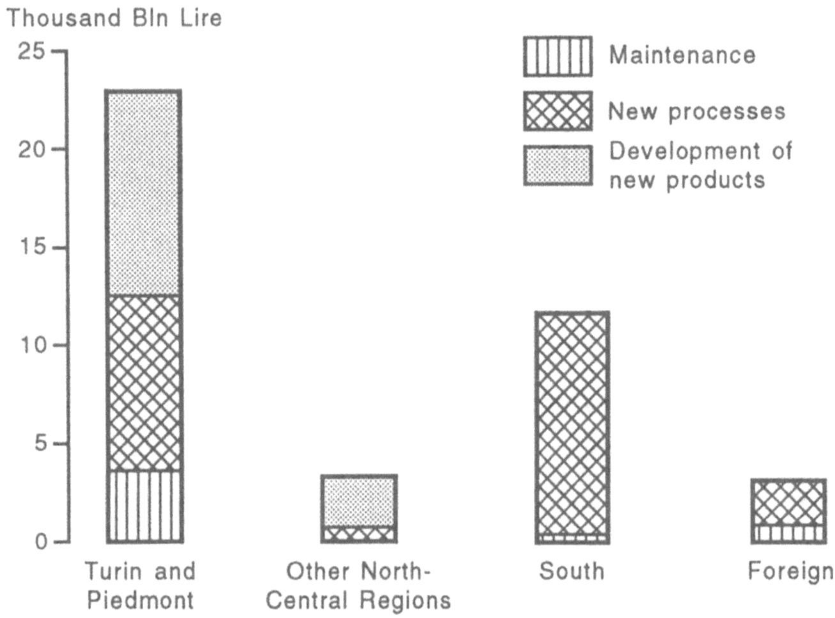

Fig. 6.3 Fiat Auto investment programmes, 1992-2001.

In conclusion, the progressive location of factories in the South - so that from 1994 more than 60 per cent of Italian production will come from this area - expresses "discontinuity" in Fiat's strategy. The position may be summarised as follows. Firstly, the policy of relocation towards the South represents a break with the past compared to Fiat's classic historical location policy, an organisational structure centralised and concentrated in Piedmont. Production is now scattered in factories distributed throughout many regions (Lazio, Molise, Abruzzo, Campania, Sicily and Basilicata). Secondly, starting in the 1980s, the southern factories, both old and new, have been the places in which Fiat Auto has experimented and introduced not only process innovations but also new models of production organisation and of industrial relations. Thirdly, until the 1970s, Fiat's production was locationally polarised, with the North constituted as an integrated system and the South specialised in only a few functions. In the 1990s, there will be a single nationally integrated model, with "sub-integration" in North and South. Fourthly, in the context of the changes just described, "continuity" is represented by the company's central functions (top-level management, research, management training, purchasing) remaining concentrated in the North. In the "historical" region of Turin, there is a strengthening of the strategic metropolitan role (management functions, R&D, marketing) and the trend towards specialisation in production with a highly innovative content. This can be deduced from analysis of investment plans made public by the company for 1992-1996[19]: investment aimed at process innovation and for maintenance of production efficiency will be concentrated in Piedmont with, respectively, 36 per cent and 52 per cent, of the company's total investments. A much higher investment share (about 80 per cent) will go to Piedmont for new products (see Figure 6.3). By 1996, 18 new models should be in production.

6.6 Concluding Comments

To sum up, the spatial perspective re-establishes its own ability to provide an holistic summary of the data on the organisational structure and the competitive dynamics which have determined the transformations of the system. What establishes itself in the 1990s is a model which distinguishes itself from previous ones, as we have seen, but this does not wipe out an automobile production system which seems, in contrast, to reproduce according to a sort of genetic code. There is, in other words, a passage in progress towards an organisational model which Butera (1991) has depicted with the metaphor "from the castle to the network", but where the castle (procedures, hierarchies, formalisations) still do not disappear but remain a structural component in the new emerging model. The Italian automobile industry, and Fiat in particular, will not become something radically different as it evolves and its trend of change will be built along the lines of its history.

Figure 6.4 simplifies and sums up the path followed, breaking down different problem areas which should not logically be separable. Right up to the end of the 1960s, the polarised model, a spatial expression of scale phenomena, was accompanied by a highly hierarchical company structure (which can be codified in the classic pattern of structure-conduct-performance) and with equally hierarchical relations within the *filière*. The major company's control clearly dominated any possible logic of collaboration, producing strongly dependent relations in the field of supply companies.

This was not destroyed in the years of concerted diffusion, which was achieved through the "duplication" of the Turin production structure in non-metropolitan Piedmont, in the Mezzogiorno, in Eastern Europe and Latin America (Brazil). The continuing strategy of quantitative expansion at that time followed an unchanged technological model (highly labour-intensive mass-production) and the decentralisation of production in Italy and abroad

[19]The overall total is 25,766 billion lire for the period 1992-1996 (and a total of 40,000 billion lire between 1992 and 2000) (information enclosed in the union agreement of 2/7/92).

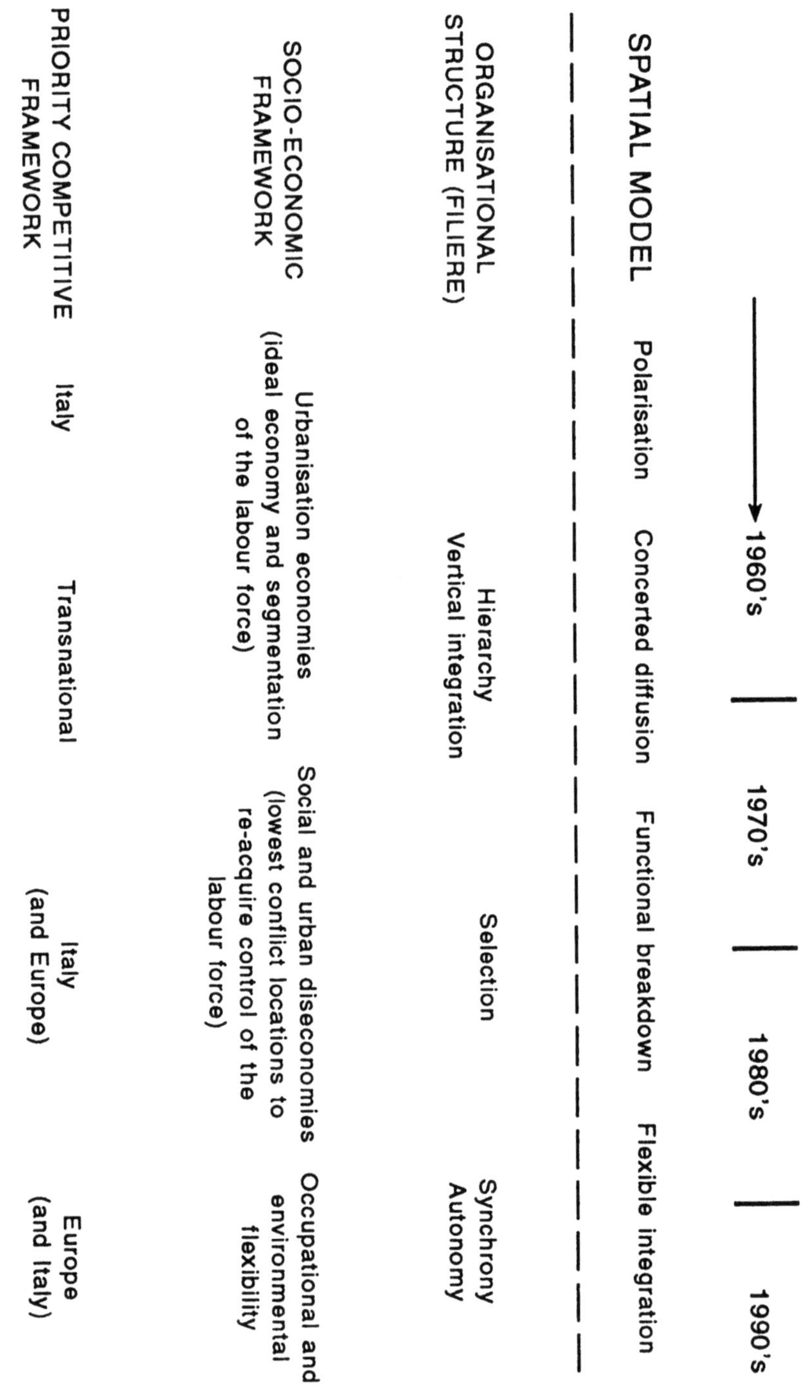

Fig. 6.4 A developmental model of the Fiat system.

followed the search for greater flexibility and lower costs in production factors. This strategic decision was to produce a timid functional subdivision in the production cycle within the system.

The reorganisation of the system of subcontracting is the fundamental strategic element which accompanies a lengthy and intense rationalisation. It was achieved initially through the selection of supplier companies, highlighting the leading units which were to be allocated autonomy in the planning and development of increasingly complex functional subsystems. Later, as we have seen, the relationship became more one of co-operation and partnership, giving the suppliers ever broadening planning functions and the organisation of synchronic forms of just-in-time production.

It is this "new" form of organisation which, from a competitive standpoint, once again gave Italy a fundamental strategic function in the changing international scenario, which has led Fiat progressively throughout the 1980s to withdraw as a direct producer of vehicles.

In contrast to its exit from relatively marginal commercial areas in terms of production volumes, Fiat adopted an internationalisation strategy based on wide-ranging agreements, such as the one with Ford Europe which aimed to establish the largest European automobile group, with almost 25 per cent of the market. Once obstacles at an industrial level had been overcome, agreement was not reached for organisational and managerial reasons: who would have had greater control over the new company? who would have led management? who would have had to make the greatest efforts in terms of management and production structures? (Pearce, 1987). In 1989 an agreement with Saab failed, as Saab preferred to ally with GM because of the wider possibilities of collaboration (cars, electronics, aviation and the military sector). The failure of a wide-ranging internationalisation strategy thus led Fiat to a reconcentration of production and its market in Italy. The low level of investment planned for abroad (lower than 5 per cent, as can be seen from Figure 6.3) seems to reiterate how the recovery (or maintenance) of market share is based on the strategic choice of great rationalisation of the Italian factories.

And yet it is also true that the international strategy has regained drive: thus, in 1992, Fiat acquired 91 per cent of the capital of FSM in Poland, set up Fatia in Algeria, of which it owns 36 per cent of the capital, and strengthened its presence in Turkey, while the realisation of the Yelabuga project in the former Soviet Union is still alive. These choices actually underline an international development along two traditional axes (eastern Europe and part of the Mediterranean), which should be linked to the company's vocation for the production of small to medium sized cars, for which production is expected to reach around 3 million vehicles per year by the end of the century.

The final question to be posed is the one concerning the use of the considerable production capacity which Fiat is setting up, in a continent characterised by production over-capacity of about two million vehicles and in an historical phase when further investment is planned by Japanese manufacturers, both in eastern Europe[20] and especially in the United Kingdom. In a market that is so competitive in terms of costs, prospects for Fiat are thus concentrated on the growth of the eastern European market and, in the west, in the ability to recover through renewal of the range. The success of this project is, however, closely linked to maintenance of the current share of 45 per cent of the Italian market. The aim, in itself defensive, is not, however, that easy to achieve in that, if all other conditions remain the same, the Japanese presence will increase in the near future, as laid down in the agreements stipulated in 1991 between the EC and Japan[21]. As it is almost certain that the Japanese share will grow, the maintenance of Fiat's share at its current level necessarily means losses by other foreign manufacturers. The complete renewal of the model range is therefore Fiat's real gamble for the future[22].

Appendix

In the traditional production structure, each Fiat plant is divided into three parts: manufacturing, which controls production and maintenance; technical services, including the planning of maintenance and technology link-ups; production services, including logistics (Figure 6.2).

With the Integrated Factory, the plants are re-organised around just two operating units: the "common services and plants" units for activities which cannot be decentralised, such as energy production, and the operational production units, which are independent of the technical and management points of view. Production units are, in turn, divided into "Production", responsible for manufacturing itself and for materials management and planning, and "Production Engineering", which ensures the effectiveness of the technical system through maintenance men and specialists and also manages the evolution of products and processes.

The basic operating structure of the production unit is the Elementary Technical Unit (ETU), which is entrusted with the government of elementary technological sub-systems, characterised by a homogeneous process and/or

[20] The countries of eastern Europe absorbed about 1,400,000 cars in 1992, with an estimate of about 2,000,000 in 1995 and 3,000,000 in the year 2000.

[21] The agreement lays down that the total share of Japanese cars in the EC can rise from the present 11 per cent to 16 per cent in 1999. Italy and France have been allowed to limit the share to 10 per cent at the end of the period. As far as Italy is concerned, this means giving the Japanese almost 100 per cent of the growth forecast for the market, maintaining the volume of European companies' share unchanged.

[22] In this scenario, one favourable condition is certainly the devaluation of the lira, which allows both recovery of internal costs, whose increase was not covered because of fixed exchange rates and by greater financial margins to be used in the investment process.

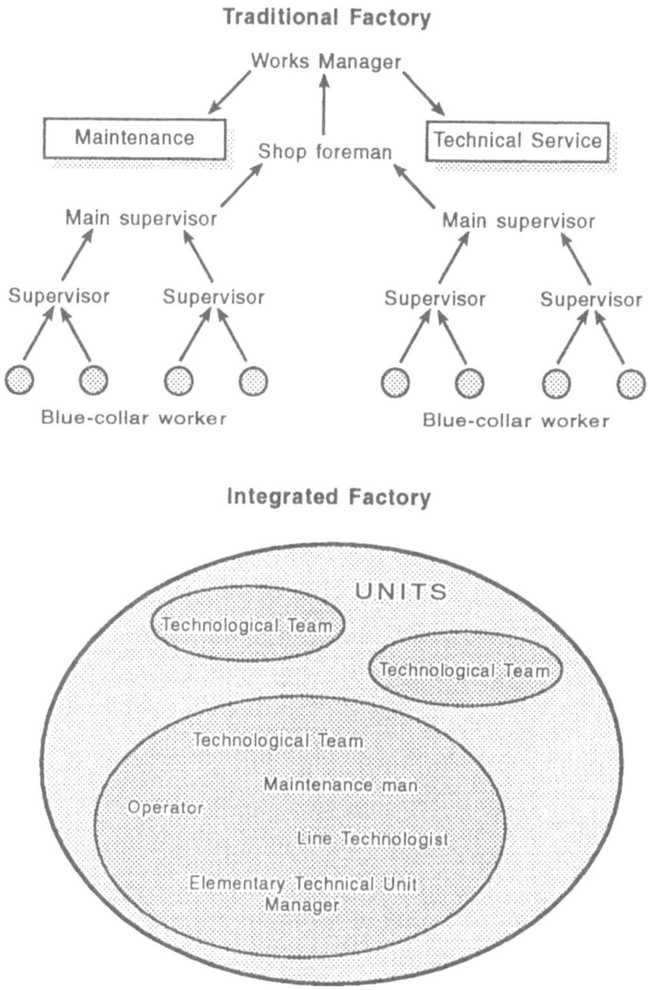

Fig. 6.5 Interfunctional integration and problems management (Source: Fiat Auto).

product (for example, in the case of engine production, the ETU manages, in addition to assembly, the production of the engine block, the driving shaft, the distributing shaft, the piston rod and the cylinder head), with the aim of optimising production processes, to improve results in terms of competitivity, quality, costs, mix and quantity. From the organisational point of view, within the ETU the number of hierarchical levels is reduced, non-hierarchical forms of organisation of work are present, such as the "technological team" (made up of the elementary technical unit manager, the line technologist, the maintenance man and operators: see Figure 6.5), and for workers, the multi-functional, regulation and process micro-management aspects are increased within a model which demands co-operation.

Chapter 7
Competitive Strategies in the World Market: The Case of Renault and the Emergence of a European Group?

Julien Savary
Laboratory for the Study of and Research on the Economics of Production, University of Social Sciences, 31042 Toulouse, France

7.1 Introduction

This chapter focuses on the recent strategies of one French national champion, the state-owned group Renault. At the same time, it reflects upon the strategies of all the main independent European car manufacturers. For they all are facing increasing competition from new Asian competitors and have tried to implement the same policy responses: first, improving their approaches to designing, manufacturing and delivering attractive new models, and, secondly, growing through internationalisation. All the groups - including Renault - have achieved some success in relation to the first objective. But all have failed in relation to the second: none became a truly global competitor; all remained heavily dependent on European markets, and often on their own national market. Renault's failure in the USA and its weaknesses in internationalisation in the 1980's are symptomatic of this wider tendency. Therefore the attempt made by Renault between 1990 and 1993 to construct a strategic alliance with Volvo can be seen as an exemplar of a new form of strategic response which could be adopted by other European groups. Indeed, three months after the failure of the Renault-Volvo merger, BMW took over Rover.

This chapter analyses in detail the Renault-Volvo alliance, because the reasons which led these groups to seek a merger reveal the basic forces which are pushing all the European manufacturers towards such close cooperation and mergers. This kind of strategic alliance currently seems the only way for the

European companies to become, initially, truly European-wide groups, and, subsequently, more truly global ones. As the collapse of the Renault-Volvo merger sharply revealed, however, such alliances have no automatic guarantee of success.

After summarizing the sectoral constraints in the automobile industry and the need for strategic changes, the first such responses of Renault will be considered: that is, improving its efficiency and increasing the international scope of its activities. Following this, Renault's attempt to find a European answer through a merger with Volvo will be considered. Finally, its future strategic prospects will be sketched out.

7.2 Why Did European Automobile Companies Have to Make Strategic Changes?

All the European (and American) car manufacturers faced the same three challenges in the 1980s (see Savary, 1994). First, competition became more acute as world demand grew more slowly, then fell, for a number of reasons: the position of automobile as a mature product in industrialised countries; the general economic crisis and the slow aggregate economic growth rates. The European market collapsed in the early 1990s, declining by 1 per cent in 1992 and no less than 15.2 per cent in 1993.

Secondly, there were important process and product innovations which led to new sectoral efficiency norms, with higher quality and lower costs and prices. This challenge, which appears to be a Japanese challenge, was vividly spelled out by Womack et al. (1990). Indeed the third challenge has been the surge of new competitors, principally the major Japanese groups. Utilizing these new ways of designing and manufacturing - "lean design" and "lean production" - the Japanese car manufacturers quickly gained important market shares in Asia, North America and Europe. In order to circumvent protectionist trade barriers, they established manufacturing units in the USA and in Europe, bringing competition into the formerly protected territories of Ford, General Motors, Renault, Peugeot, Fiat, Volkswagen and others.

Lastly, deepening European integration implies that the automobile national markets, which to date remain relatively protected, could be rather quickly opened to Japanese competition (see Hudson, this volume). This removal of national and European market protection, which seems inevitable in the medium term, will create an important challenge for European automobile manufacturers. In 1987 the Japanese only held 1 per cent of the Italian market, 15 per cent of the German market, 11 per cent of the British market, 4.5 per cent of the French and 3 per cent of the Spanish market (Sadler, 1992b). That will change: it is claimed, for example, that "Japan's market share in Europe could double, to 22 per cent by decade's end" (Business Week, 23 November 1992).

Faced with these three challenges, automobile producers in Europe have evolved two strategic responses. All the major automobile groups - both the Americans (Ford, General Motors and Chrysler) and the Europeans (Renault, Peugeot, Fiat, Volkswagen, Volvo, Rover, Mercedes-Benz) have implemented the first of these strategic responses: that is, they have changed their approach to carrying out research, development and manufacturing, in order to attain Japanese norms, increase the quality of vehicles and reduce their cost. This involves, inter alia (Womack et al., 1990): reducing the time taken in designing new models; increasing internal communication and rapid adaptation to consumer demand; improving product quality at every stage of the value chain; reducing manufacturing costs through automation of production and just-in-time organisation; and instigating a new supply policy, with fewer and more competitive suppliers.

The European car manufacturers also developed international cooperation agreements as a means of increasing efficiency. Agreements with Japanese firms as a route to learning about lean production were rare: the main one was between Rover and Honda (see Hudson, this volume). European agreements in search of scale economies have been more frequent; for example, that between Renault and Peugeot in 1966, or that reached between Renault and Volvo in 1990. The European groups have also implemented international agreements in order to enlarge their product range, with new specialised vehicles. These agreements have included both distribution and manufacturing. Such agreements were also often forged in relation to cooperative research, to develop either key components such as engines or transmissions or new car models.

International growth through exports and direct investments was another strategic response, deployed by all the European groups, in order to increase their size, reduce costs and improve competitiveness. Despite this, in the 1990's, the non-globalisation of the European firms is very apparent. Their vehicle assembly operations outside Europe range from 3 per cent (PSA), to 5 per cent (Renault), 10 per cent (Fiat) and 19 per cent (Volkswagen). Volkswagen, with important activities in South America and in Mexico, is the most globalised group, but its activities in North America and in Asia are weak.

The four European generalist car manufacturers are little internationalised. More than half of their total production volume is carried out inside their own home country: 55 per cent for Renault, 63 per cent for Volkswagen, 80 per cent for PSA and Fiat. Insofar as they have internationalised, it is mainly in the EC, largely in Spain and Portugal. All have implemented a strategy of decentralizing production into southern Europe. Controlling manufacturing operations in Portugal and Spain has been one way of accessing these protected markets while reducing costs. They also manage assembly operations in Turkey and did so in the former Yugoslavia. In Eastern Europe they also developed small scale cooperative agreements: Fiat was the most heavily involved in the USSR and Poland (see Conti and Enrietti, this volume). The opening up of these countries recently led to some new investments, such as Volkswagen's acquisitions in the Czech and Slovak Republics. In North America the European groups remain very weak: for example, in the late 1970s, Volkswagen had opened a plant in the USA, but later closed it (although it is still managing an assembly plant in Mexico, within the new North American Free Trade Area).

In the 1990's European car manufacturers have yet to become substantially internationalised: "most European car manufacturers appear to be concentrating on their own region Europe, with a local (PSA, Fiat, Mercedes-Benz) or regional (Renault, Volkswagen) production basis, and often a regional market basis (PSA, Fiat, Renault)" (Tulder and Ruigrok, 1993, 28-9). In contrast, their competitors are truly global or "multiregional" firms with important investments in the three triad areas of Europe, North America and south east Asia. Ford and General Motors have strong subsidiaries and hold significant market shares in Europe, and are present in Asia through international alliances: Ford with Mazda and Kia, GM with Isuzu and Daewoo, while Chrysler has links with Mitsubishi and Hyundai. The Japanese groups also have important investments in North America, and growing factories and market shares in western Europe (see Hudson, this volume). This difference could push the European groups to accelerate their international growth through intra-European cooperations and mergers. The Renault case exemplifies these trends.

7.3 A Basic Answer : Improving Efficiency

The first strategic response adapted by Renault, like other European groups, has been, for fifteen years, to invest in new plant and machinery and develop new management methods in order to reach the efficiency levels of the Japanese, which have become the new sectoral norms. There have been a number of dimensions to this form of response.

7.3.1 Project structures

Renault has restructured its internal organization in order to increase flexibility, adaptability to changing demand, and enhance competitiveness. In the 1990's it decided to improve the efficiency of its internal structures, by reinforcing the project groups for the management of new model developments. These groups had first been created in 1975. This meant that an increasing role was given to these project teams and more emphasis placed upon the process of parallel development, new approaches to human resource management in the light of changing project tasks, and new relations with suppliers based on partnership in jointly developing new methods and products (see Midler, 1991). The aim was to develop more efficient forms of internal communication and cooperation, imitating the Japanese "heavy project groups" model. This policy enabled Renault to improve its design and development procedures and reduce the total time necessary to develop a new model: the development of the R19 took more than 4 years, at a time when the Japanese producers needed (on average) 3 years (Les Echos, 5 December 1992). But this was lowered for the most recent models, the Clio and Twingo.

7.3.2 Development of quality

Renault also improved the quality of work at all the stages of the value chain, from R&D to manufacturing, delivery and after sales services. Introducing other innovations, such as changes to projects structures, automation, and just-in-time manufacturing, enabled this quality improvement. But it also implied training of workers, and new practices such as regular meetings of all employees about quality. Indeed, in recent years Renault product quality appears to have improved to competitive levels, and an independent survey by the German's motorists' association even put the quality of the Renault 19 ahead of the Volkswagen Golf (Economist, 5 December 1992). The more recent models, like the Clio, the Safrane and the Twingo are also of high quality.

7.3.3 Automation, just-in-time manufacturing and supply

Renault is also cutting costs in manufacturing, through plant automation. The number of robots doubled in four years from 1985 to 1989 (see Table 7.1). Automation, mainly of the most dangerous and arduous tasks like painting and welding, allowed productivity to rise. There are, however, more recent trends to reduce the level of automation: automated welding work stations fell from 98 per cent for the Clio to 70 per cent for the Twingo (Les Echos, 11 January 1994), reducing investment costs in the new plant by 20 per cent to 30 per cent (Les Echos, 21 March 1993).

At the same time new ways of working are being introduced. Operators have become less specialised, and also play a role in quality control. This job enrichment implies extra training costs and changes in levels of qualification. Adopting the just-in-time model in the organisation of internal and external operations also reduced stock volumes and costs, and quickened the response to demand changes.

The consequence of improved work efficiency, at a time when the total production volumes increase slowly (as they did until 1989) and even decrease, is a corresponding reduction of the workforce. Renault vigorously pursued such

Table 7.1 Trends in Renault group operations

	1985	1986	1987	1988	1989	1990	1991	1992	1993
Unit sales (thousands)	1,717	1,816	1,901	1,913	2,003	1,871	1,882	2,045[1]	1,800
Workforce (thousands)[2]	196.4	196.7	188.9	178.7	174.6	157.4	152.0	146.6	
Number of suppliers			960	900	810	720	680	630	
Number of robots[3]	453	459	620	828	890				
Market share[4] (per cent)									
in France	28.7	31.5	30.5	29.0	29.2	27.9	26.8	29.5	30.6
in Europe	10.7	10.7	10.6	10.2	10.4	9.9	10.0	10.6	10.5

[1]not comparable to 1991

[2]in 1990, 1991 and 1992 Renault Argentina is excluded (8,700 employees in 1989), in 1992 Oyak Renault is added (4,932 employees)

[3]5 or more axis numerical control robots

[4]Renault's market share (passenger car market)

Sources: Renault Economic Atlas, 1993; Renault Annual Report,1992; various newspapers

a decrease and reduced its total employment by about 67,000 people between 1984 and 1992 (see Table 7. 1). This trend will continue as European demand remains depressed and the efficiency level of Renault workers still lags behind Japanese norms. Renault S.A., which had cut employment by 2,500 during the year 1993, intended to further reduce it by 2,350 in 1994 (Les Echos, 21 December 1993).

Average productivity per employee has therefore risen strongly. Between 1991 and 1992 the number of vehicles per annum manufactured by each employee increased from 12.15 to 14.28 (see Table 7.2). This productivity growth is also reflected in the reduction of the "cycle time", that is to say the average time spent by each worker in performing his or her allocated

Table 7.2 Renault group productivity growth between 1991 and 1992

	1991	1992
World production (number of vehicles)	1,848,078	2,094,774
Total employment	152,050	146,604
Number of vehicle by employee	12.15	14.28
Increase in productivity, 1991-1992		+ 17.6 per cent

Sources: Renault Annual reports

operations on the vehicle. This fell from 12 minutes in the 1960's to 1 minute in 1994 (Les Echos, 11 January 1994).

7.3.4 A new supply policy

External supplying has been increased so that the degree of internal supply within Renault decreased from 31 per cent in 1978 to 19 per cent in 1985. Renault also sold several of its components subsidiaries, such as Renix to Allied Signal (electronic components) and Les Peintures du Mans to Herberts (vehicle painting). This trend towards increasing sub-contracting and external supply continues. For example, the braces which will be used for the diesel engine of the new Laguna model will be manufactured outside the Renault plants, and supplied by its subcontractor CF2M. Formerly supply of this key component was subcontracted only in short runs (Les Echos, 21 January 1994).

This new supply policy has been implemented with fewer first-rank suppliers. Many of these are Renault's sole suppliers of particular components. These suppliers are then more closely involved in Renault's technological and organisational developments. Renault has not only reduced the number of its suppliers by more than one third during the last six years (see Table 7.1) but it has also increasingly concentrated on particular suppliers for particular components, thereby creating an integrated European network of suppliers.

This new policy placed all potential Renault suppliers in competition at European (even global) level. It has not always led to these suppliers locating close to Renault plants. On the contrary, for low value and standardised parts, locational considerations such as longer working hours and lower wages are influential. For one given plant, this may enlarge the geographic area of sourcing.

Only a few large first-rank subcontractors are located close to the plants, to deliver just-in-time low value and high volume parts, such as seats, mufflers, bumpers, instrument panels and other plastic parts. These subcontractors are also often themselves subcontracting numerous parts, which they are assembling into sub-systems. For instance, a few manufacturing suppliers are located near the Renault assembly plant of Sandouville; for example, Reydel (consoles), Plastic Omnium (bumpers and other plastic parts) and Imetal (mufflers). Other subcontractors have located warehouses close to the plant for buffers stocks; for instance, transportation companies such as Vimbert, which is delivering all the tyres. In 1991, Tubauto and Roth Frères created a new plant close to Sandouville, to deliver the seats of the new car which was the successor to the Renault 21 (Les Echos, 23 October 1990). But of an FF 8-9 billion of total purchases, only 13 per cent are contracted to 22 local suppliers, located less than 35 minutes from the Sandouville plant. These suppliers are called "synchronous" suppliers (Les Echos, 12 February 1992). These trends seem to be similar to those existing for German car manufacturers in Germany (see Schamp, 1990 and this volume; Gaebe, 1991b).

7.3.5 Restructuring the European plants

Another route to cost cutting has been a plan to restructure all the European units, which was put together in the 1980's. The aims of this were to improve the functional specialisation of each plant, increase production volumes to benefit more from scale economies at each plant, and reduce the number of plants. The plan argued that "to make the most efficient use of its production facilities, Renault has no other choice but to reduce the number of plants in operation, and to concentrate assembling at its most cost-effective sites".

Some of the main features of that plan were:

(a) The phasing out of Billancourt (France) plant, announced on November 21st 1989. This old plant had been modernised, and its productivity had increased but it remained below that of other plants. Furthermore, its

Table 7.3 Renault's main European plants, 1992

Plant	Product	Annual Production
Cléon (F)	Engines	886,151
	Gearboxes	1,011,412
Le Mans (F)	Casting and axle assemblies	1,194,554
Romorantin (Matra plant) (F)	Renault Espace	55,355
Creil (F)	Renault Trafic	43,299
Douai (F)	Body assembly, R 19	316,989
Flins (F)	Body assembly, Renault Clio	363,865
Sandouville (F)	Body assembly, R 25, R21, R Safrane	173,456
Maubeuge (F)	Body assembly, R 19, R Express	117,236
Batilly (F)	Renault Trafic/Master, Gamme B	28,384
Vilvoorde (B)	Body assembly, R 21, R Clio, R Nevada	189,529
Valladolid (S)	Engines	348,529
	Body assembly, R Clio, R Express	195,394
Palencia (S)	Body assembly, R 19, R 21	168,967
Sevilla (S)	Gearboxes	568,811
Cacia (P)	Engines	193,314
	Gearboxes	136,746
Setubal (P)	Body assembly, R S 5, R Clio	72,808
Novo Mesto (Y)	Assembly, R4, R Super 5	83,353
Bursa (T)	Assembly, R 9, R 11, R 12, R 21,	110,644

F France; S Spain; B Belgium; P Portugal; Y Yugoslavia; T Turkey
Source : Renault Economic Atlas - 1993

location, on an island in the Seine, prevented both the extension of the plant and the location of subcontractors near to it. The 1989 production of 107,000 units was far below the minimal efficient size in assembling, which was of more than 200,000 units. So production was shifted to other Spanish and French plants, such as Maubeuge and Flins. The plant was closed in 1992.

(b) The Guarda assembly plant in Portugal (which only assembled 7,000 Trafic vans in 1989) was sold to the German group Reinshagen, an international specialist in wire manufacturing. But important investments were made in the two other Portuguese plants. The Cacia plant manufactured 193,000 engines in 1992 and in 1993 produced all the engines for the Twingo model, assembled in Spain and in France.

(c) The Fasa-Renault activities in Spain have been restructured, and Renault shut down the older unit of the two assembly facilities - the number 1 plant in Valladolid - in mid 1991. Production has been shifted to the two other assembly plants in Spain, at Valladolid and at Palencia. This has increased their efficiency. Indeed today the minimum efficient size for an assembly plant is, at least, around 900 to 1,000 vehicles a day; that is, 207,000 to 230,000 vehicles annually. In 1989 production at the Palencia plant was 139,000 units. By 1991 it reached 152,000, and in 1992 169,000 (see Table 7.3). In 1993 the assembly plant at Valladolid (producing the Clio and Express) is Renault's most modern and recent plant. Renault is planning to invest FF 3 billion in the next five years to

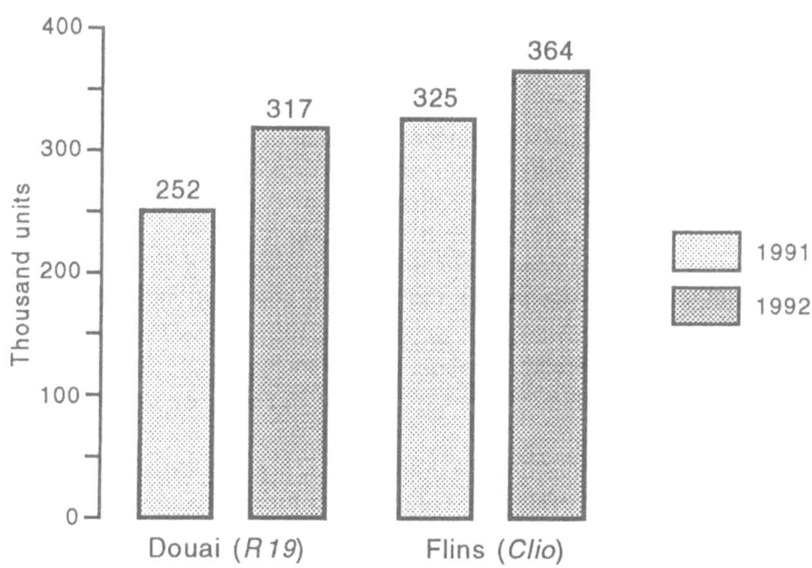

Fig. 7.1 Plant production volumes in 1991 and 1992 (in thousand units).

manufacture diesel engines in Valladolid, to be delivered to all the group (Les Echos, 4 January, 1993).

In the past, it was planned that several French plants would increase their production volumes. As a result, the Maubeuge body-assembly plant received investments, the Flins body-assembly plant was prepared to increase production through the introduction of a third shift of workers, therefore allowing more intensive use of machinery. But, because demand slackened, production did not increase in 1990 or 1991. In 1992 increased total sales and production volumes - partly because of successful new models launched by Renault - allowed important increases of production volumes in each European plant. The trends for Douai and Flins plants are shown in Figure 7.1.

In 1992 the production volumes of the European plants (see Table 7.3) were higher than those five years earlier. They were approaching the minimum efficient size, which itself seems to be rising in the automobile industry.

Another trend has been to tighten the links between Portuguese and Spanish plants, and between the commercial networks in these two countries. Renault has increased the level of integration and of globalisation of all its European activities. One significant development has been the setting up of a computerised information network linking its European plants, and also linking it to 180 of its suppliers in Europe.

7.3.6 National and international agreements

Another approach to achieving efficient sizes in the development and manufacturing of products has been via cooperation agreements with other car manufacturers. In the 1960's and the 1980's, Renault made several alliances with competitors, in order to share R&D or the costs of manufacturing components and specialised vehicles. The main agreements were signed with the French group Peugeot in the period 1966-1970, because the two groups wanted to share costs and were convinced of the necessity of cooperation. These agreements were enlarged to include Volvo in 1971. Three joint subsidiaries have been created and are still in activity: Française de Mécanique, for engines production (1,231,124 units in 1992, of which Renault took 279,600); Société de Transmissions Automatiques, for mechanical components production (gearboxes: 52,989 units for Renault in 1992); Société des Usines Chausson, for assembling Light Commercial Vehicles: 43,299 Renault-Trafic vans assembled in 1992.

This cooperation only increased until 1975, because in 1973 Peugeot took over Citroën and then wanted to rationalize its operations, mainly with Citroën (see Loubet, 1990, 351-84). Since this date, cooperation has been restricted to these three joint subsidiaries. In the 1990's, when Société des Usines Chausson faced huge difficulties, the two groups refused to invest to avoid business failure. Bankruptcy occurred in September 1993, and it seems that Renault alone will manage this company and its two plants for body manufacturing (in Gcnevilliers) and assembling its Trafic vans (in Creil). Nevertheless in 1993 the two groups decided to deepen their cooperation. They will jointly design and manufacture a new automatic gearbox for their low and middle range models.

Since 1983, with the French group Matra, Renault has been producing the successful multi-purpose vehicle, the Renault-Espace. This model was designed jointly, but is manufactured at the Matra plant of Romorantin (55,355 units in 1992), and sold through the Renault sales network. This agreement has been recently extended until 2000. The European sales network of Renault had also been distributing Chrysler Jeeps since 1987, but that agreement ended in 1993, and Renault is now looking for another supplier of four wheel drive (4WD) cars.

With Volkswagen, Renault decided in 1983 to develop an automatic transmission. Cooperation with DAF, to develop a new range of light commercial vehicles (to replace the old Renault Master model), started in 1989 but failed in 1992 because DAF withdrew. Renault then tried to continue this project with Volkswagen. In 1992 an agreement was signed with BMW to recycle wrecks in France and Germany.

Renault is also carrying out long term research on electric automobiles with PSA. In 1992 it signed an agreement with PSA and EDF to promote these zero pollution urban cars, and determine the necessary technology. An EC Eureka research project on the "electric urban car" was launched in 1992, with Renault as a member.

Finally, the international agreements signed by Renault with non-French manufacturers are very few, and those with Japanese groups are quite absent. Outside France, Renault is assembling its car models in several less developed countries, through agreements with local manufacturers (see below) . In addition, in Colombia it is assembling Toyota 4WD models in its Medelin plant.

7.4 A Missed Opportunity: Renault's Failure to Grow Outside Europe

A second strategic response has been to increase the extent of its operations outside Europe, through exports, acquisitions and minority participations into foreign manufacturers, though such activity has been quite limited in the last ten years. Renault remained a European-based group, with strong operations in Spain, Belgium, (former) Yugoslavia and Turkey, and some assembling agreements in some parts of the Third World.

7.4.1 The missed deal in the USA

Renault failed to establish itself in North America, despite vigorous efforts. In 1982, it subscribed to $40 million of convertible debentures in the American Motors Corporation (AMC), in which it had had a minority holding since 1979. With 46.4 per cent of the capital and technical agreements signed, Renault had virtual control over the fourth biggest American car manufacturer, employing 146,000 workers and with a turnover of $3 billion. AMC distributed the R5 and R18 in the USA, assembled the Fuego model, and in September 1982 began local production of the R9, christened the Alliance, with a target of 100,000 sales per year. In 1983 Renault invested a further 50 million dollars for AMC to produce the American version of the R11. It also took local control of two companies in Mexico, which were already assembling Renault and AMC cars. In 1984, it decided to create a new plant in Mexico to produce engines, for export to the USA so as to reduce production costs for the Alliance and other AMC models. In March 1987 AMC began to sell imported Renault 21 cars, called "Medaillon", and in the summer of that year began to produce the new midsize "Premier" with Renault imported parts, in Toronto.

But Renault faced huge difficulties both in Europe and in the USA, where sales grew more slowly than anticipated. To cut losses, Renault introduced a restructuring plan from 1985, cutting employment and selling several plants in France (including Renix, Bernard Moteurs and Renault Machine Outil) and abroad (for example, in 1985 it sold its stake in Volvo). In 1986, AMC closed its Mexican subsidiary which manufactured jeeps. Lastly, in early 1987, Renault sold its 46.4 per cent stake in AMC to Chrysler (paid by $200 million in notes and $35 million in cash). After having heavily invested in American Motors -$645 million in seven years- it sold this company in the year in which it again reached profitability. This sale showed the failure of an ambitious plan - Renault was the only European group which tried to take over an American car producer - mainly due to increasing competition within the USA market, the weakness of AMC which was specialised in the less profitable small car segment, and to strategic constraints on Renault.

In 1979 Renault had bought 20 per cent of the capital of Mack Truck, a USA heavy lorry producer. In May 1983 Renault doubled its stake, taking responsibility for management. Since 1990, it has owned 100 per cent of the capital of this company, which holds 13 per cent of the USA market for trucks of over 15 tons. This success made Renault the second biggest producer in the world of utility vehicles.

7.4.2 The weak positions in other countries

In Africa, Renault is mainly an importer, only assembling 6,900 cars a year in Morocco. In the Middle East, Renault has invested in Iran since the late 1970s, where a state-owned company, SAIPA, assembled cars with imported kits

(9,347 cars in 1992, 22,000 in 1991). But this cooperation may stop, because SAIPA chose the Korean company Kia for a new assembling agreement in 1993.

In Asia, Renault is absent from several major markets, notably Japan (in which it sold only 1,800 cars in 1992), India, and Korea. Its main Asian market is in Taiwan, where a local company assembles 16,000 Renault cars a year. Renault also produces cars in Volvo plants in Thailand and Malaysia, as a result of cooperation with Volvo. but on a very small scale (1,000 cars a year). In 1993 Renault first invested in China: a local joint-venture will assemble light Renault-Trafic trucks, converted to minibuses (40,000 units a year in 1997).

In Latin America, Renault sold 93,816 cars, but its market penetration varies greatly from one country to another. It is absent in the biggest market of Brazil. In Mexico, it formerly manufactured R9 and R18 models (19,000 units in 1985), until it closed its plant in 1986 in order to reduce the financial losses of the group. Its subsidiary RIMEX (Renault Industria de Mexico) was created in 1984 to manufacturing engines delivered to American Motors in the USA. After the withdrawal of Renault from AMC, RIMEX continued to produce engines, which are now delivered to the European plants of Renault. With 1,160 employees and a production volume of 252,081 units, it is one of Renault's four engines plants, and a specialised unit in its world manufacturing system.

In other Latin American countries, Renault has reduced its financial investments in local companies. In Argentina, it manufactured cars in a majority owned subsidiary until it sold the major part of its capital stake in that company in 1991. But 70,000 Renault cars are still assembled each year in Argentina, and Renault controls 25 per cent of the car market. In Colombia, it was managing a subsidiary, Sofasa-Renault, until it also sold its majority participation in 1993. Sofasa will continue to assemble Renault cars (the R9 and R21 - 6,700 in 1992) and produce the R19 model. Renault also manages a subsidiary in Venezuela (assembling 6,700 cars a year), but has sold its Chilean subsidiary.

In Eastern Europe, Renault had several cooperation agreements originating from the 1950's (see Badiche, 1992). It then tried to increase its investments following the political changes of the 1990's. But in fact, by mid-1993, only a few investment projects had been launched, and a major failure occured. For many years Renault, using imported CKD kits, had been producing cars in Yugoslavia, which were exported to Europe. In 1990 it was planned to double the volume of this production to 75,000 units in 1991, but the recent civil war led to the cancellation of this project. The sales of Renault are increasing quickly in Germany, and Renault is setting up a commercial network in its eastern regions. Renault is discussing proposals with its previous associate Dacia in Rumania (see Maxcy, 1981), to create a new production capacity of

120,000 units per year (see Chanaron, 1992). In the same country a project for an assembly plant with Olcit (for light commercial vehicles and the R5 model) was stopped in 1993, because the government refused to protect the national market from imports (Les Echos, 22 September 1993).

Renault's main proposed projects in eastern Europe were in Czechoslovakia, but it again failed, twice: in September 1990, to cooperate with the local group Skoda, which chose Volkswagen instead; and in 1991, with another local partner Avia in a small trucks project. Renault also had other projects there: to assemble 15,000 Trafic vans with BAZ; and for an engine plant with ZTS. It successfully concluded a joint-venture with the local firm Karosa, to modernise and retool a bus factory (Les Echos, 11 January 1993).

In conclusion, Renault's total production volume outside Europe is meagre: only 319,853 vehicles in 1992, some 15.3 per cent of its world production, and a figure corresponding to the production of only one efficient assembly plant, such as that at Douai. Turkey (100,000 R9, R12 and R21), Slovenia (83,000 R4 and R5) and Argentina are the main foreign countries for manufacturing (see Table 7.4). Similarly, Renault's sales outside Europe also represent only 11.3 per cent of its total sales (see Table 7.5).

Table 7.4 Renault's world production by geographic area in 1992 (automobiles, light trucks and trucks)

	Units	Per cent
France (auto)	1,110,614	
France (trucks)	33,701	
Total France	1,144,315	54.6
United Kingdom (trucks)	1,185	
Spain (auto)	364,361	
(trucks)	2,723	
Belgium (auto)	189,529	
Portugal (auto)	72,808	
Total other Europe	630,606	30.1
Total Europe (France included)	1,774,921	84.7
North America (Mac Trucks)	15,336	0.7
Latin America	79,407	3.8
Turkey	110,644	
Asia (Taiwan, Iran, Thailand, Malaysia)	24,209	
Morocco	6,904	
Slovenia	83,353	
Total Asia Middle East Africa	225,110	10.7
Total (world production)	2,094,774	100.0

Source: Renault Economic Atlas 1993

Table 7.5 Renault turnover by geographic area 1990-1992
(Renault group sales by country in per cent of total sales)

	1990	1991	1992
France	51.1	48.3	47.7
Other Western Europe	37.0	40.7	39.2
Eastern Europe	1.1	1.4	1.8
Total Europe	89.2	90.4	88.7
Africa	2.5	2.0	1.7
Americas	6.2	4.9	5.5
Asia-Pacific	2.1	2.7	4.1
Total	100.0	100.0	100.0

Sources: Renault annual reports

7.5 The Result: The Weaknesses of Renault in the 1990s

In 1990-1991 Renault appeared as a more competitive group, but one which persistently lagged behind its Japanese competitors in competitiveness and degree of internationalisation. Its financial results were nonetheless improving: net earnings rose, and the growth of indebtedness slowed (see Table 7.6). But analysts still thought that its indebtedness was still too high: "Yet the burden is

Table 7.6 Trends in Renault group financial results
(in billion of FF or per cent)

	1987	1988	1989	1990	1991	1992
Turnover[1]	147.5	161.4	174.5	163.6	166.0	179.4
per cent abroad	51.3	51.8	50.8	48.9	51.7	52.3
Net earnings (Renault's share of consolidated earnings)	3.3	8.8	9.3	1.21	3.08	5.68
Net financial indebtedness on December 31st	46.4	23.8	17.6	27.1	15.5	12.5
as a percent of turnover	31.4	14.7	10.1	16.6	9.4	7.0
Total capital expenditures[2]	7.0	7.3	10.4	13.2	21.6	11.7
of which financial investments			-0.6	2.54	12.1[3]	0.5

[1] financial subsidiaries excluded
[2] property, plant and equipment, intangibles, financial investments
[3] capital stake in Volvo companies
Sources: 1992 Renault annual report

Table 7.7 Japanese, European and Renault manufacturing performances

	Japanese groups (end of the 80's)	European groups (end of the 80's)	Renault (in its most efficient plants in 1990)
Time period for designing a new model	46 months	57 months	54 months for Clio
Number of suppliers	170	442	230 for Flins plant
Time for final assembling of one car	17 hours	35 hours	19 hours for Clio model
Employees working in teams (in per cent)	69	6	80 per cent in Flins plant
Training time by employee for a new model	300 hours	150 hours	450 hours for Clio

Source: For Japanese and European average data: MIT report; for Renault: interview in "Le Nouvel Observateur" 13 December 1990

still too heavy to allow Renault to build new plants to fight the Japanese, as Volkswagen and Fiat have done" (Business Week, 23 November 1992).

Labour productivity at Renault leapt by 50 per cent between 1985 and 1991, and rose by 17.6 per cent during the year 1992 (see Table 7.2). The new just-in-time delivery schedules "have cut inventories from 10 days to 2" (Business Week, 23 November 1992). Yet Renault's managers think that its costs are still 30 per cent above the most efficient Japanese factory outside Japan, Honda's benchmark plant in Marysville, Ohio (Business Week, 23 November 1992). Even in its most modern and efficient plants, Renault still lags behind Japanese efficiency norms. The director of the Flins factory, Michel Auroy, explained (in an interview with the author in December 1990) that Renault productivity had improved in several fields, and was approaching - but was still below - Japanese norms (see Table 7.7). He said: "in 1990 the Renault Clio is assembled in 19 hours, and we fight to assemble it in 16 hours. A Japanese group could produce it in 14 hours. The gap still exists".

A recent report from the Boston Consulting Group showed that the productivity of European suppliers of automobile components also lags behind - by up to half - that of their Japanese competitors. Even if the French suppliers seem to be the most efficient in Europe, this is still another source of difficulty for Renault (Les Echos, 19 October 1993).

Furthermore, Renault remains a largely national company, especially when compared with the USA and Japanese sectoral leaders which are truly global triad-based groups. Renault's sales are still focused on France (48 per cent) and on the EC (86 per cent). They are very low in North and South America and in Asia (see Figure 7.2). Renault's car production is also mainly located in four

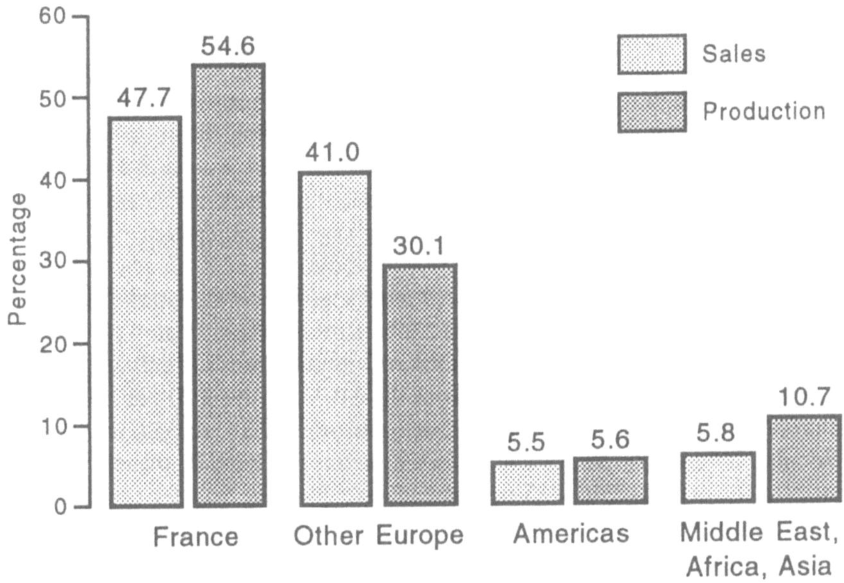

Fig. 7.2 Renault sales and production by area (1992).

European countries: France, Spain, Portugal and Belgium, and it is growing in Turkey, where the group has recently been investing (see Figure 7.2).

In order to share research, development and manufacturing costs, and develop a more international structure in the 1990s, Renault needed to grow more quickly. Its 1990 alliance with Volvo (see also Malmberg, 1991 and this volume) was, therefore, then clearly developed as a response to its perceived deficiencies, and as a means of reducing costs and accelerating its international growth.

7.6 A European Answer: Rise and Fall of the Renault-Volvo Merger

7.6.1 1990-1993: the deepening of the alliance

The French state-owned group Renault and the Swedish privately-owned group Volvo AB began cooperating in 1979: the agreement included Renault taking a minority equity stake (10 per cent) in Volvo's capital (which it has since sold). Renault has delivered engines and gearboxes to Volvo. Since 1981, Volvo has distributed Renault models in Scandinavia, and since 1986 it has manufactured them in its Thailand subsidiary.

On September 27th 1990, after months of preparation, Renault and Volvo signed a more important industrial and financial alliance concerning cars and trucks. The major aims of this alliance were to improve competitiveness and reduce costs through economies of scale and of scope in all corporate functions: R&D, manufacturing, purchasing, distribution. More precisely this alliance would allow Volvo, which was of medium size and focused on specialised products, to survive as an independent producer by obtaining access to middle market range products. It would allow Renault, which was looking for more important markets and more profitable products, to expand in the northern Europe area and in the upper product segments of the market.

The importance of this alliance was soon further demonstrated. In January 1991 the two parent companies took minority equity stakes in each other and in those of their main subsidiaries: as a result, AB Volvo held 20 per cent of the Renault's capital, and 45 per cent of the capital of its subsidiary Renault Vehicules Industriels (trucks and buses). Conversely, Renault owned 8.24 per cent of AB Volvo's capital, 25 per cent of its subsidiary Volvo Car (cars), and 45 per cent of its subsidiary Volvo Trucks (trucks and buses) (see Figure 7.3).

This alliance was developed more rapidly in relation to cars than to trucks, where the similarities between Renault and Volvo, and therefore the competition between the partners, was keener. In terms of trucks and buses, cooperation began through common projects in research, development, and purchasing. The cooperation in purchasing and in the use of common components was planned to save each company SKr one billion in truck production in 1996 (Enjeux Les Echos, March 1993). In the car sector Renault

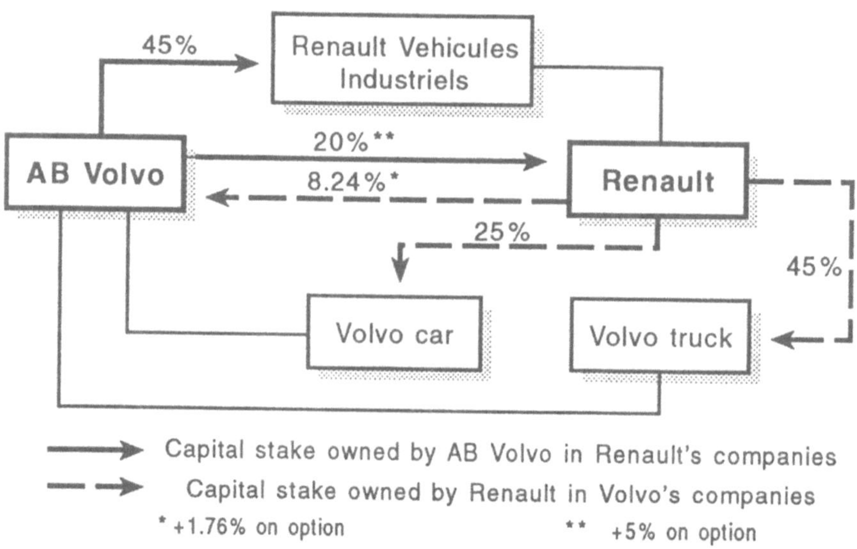

Fig. 7.3 Renault-Volvo cross shareholdings, 1991 (as of December 31st 1991).

and Volvo progressively coordinated their business policies. By the end of 1993, this cooperation embraced all corporate functions, through several "co-ordination committees" working under the authority of a "joint general policy committee", co-directed by the chairmen of Renault and Volvo. Some of these areas of cooperation will now be described in more detail:

Design and quality policies: The two groups created a joint design committee to analyse and bring together their design policies. They also began to harmonize their two internal and external quality policies: product quality; process qualities in development and manufacturing.

Research and development: In 1990 the two groups established a joint venture (ARA Associates in Advanced Research) to coordinate their future advanced research projects. By 1993, one third of all their research projects were already jointly managed. Renault and Volvo were also exchanging their skills, and sharing tasks, in the development of platforms, engines, and gearboxes, as well as in security and limiting environmental impacts of vehicle production and use. In Autumn 1992, they decided to develop a common platform for their next top-of-range models, the successors to Renault's Safrane and Volvo's 900 model. This choice represented a key move towards a merger, because it implied that these high quality products would be designed, from the outset, on a common basis. After April 1993 they also cooperated in the development of electrical and mixed engines.

Manufacturing and purchasing policies: Volvo and Renault agreed upon a 10-12 years plan for joint product development: the two ranges of models would continue to exist separately, but common platforms and components would be increasingly used, in order to avoid duplication and realize scale economies, both in R&D and in manufacturing. Important exchanges of components also began. So, from 1991-1992, Renault delivered about 100,000 engines (F model and G model) and 100,000 gearboxes annually to Volvo, for its 400 model. Volvo also delivered each year 16,000 car bodies (floors) for Renault's Clio model. In 1994 it was planned that Volvo would deliver 100,000 new petrol engines "N" to Renault. These engines were to be used in the new Renault "Laguna", the replacement for the Renault 21. Renault saved FF one billion in 1991 through purchasing the Volvo "N" engine".

In June 1993 a joint product-planning department was created, in order to facilitate these developments. For instance it was decided that, in the future, Volvo would be responsible for developing the petrol engines of the two partners, and Renault for developing the diesel engines.

Renault and Volvo also implemented a common purchasing policy. In 1992 the two groups were already jointly purchasing 20 per cent of their total external supplies (FF 17 billion out of FF 85 billion), and they were planning to increase this to 80 per cent in a few years. For these key materials and components, one sole buyer was managing the common purchases.

Marketing: The two groups started to coordinate their sales policies and operations, while strictly maintaining their different brands, ranges of products and sales networks. Each group started to distribute the others' cars in countries where it was in the strongest position. Thus Renault sold Volvo cars in Argentina and Slovenia while Volvo continued to sell Renault cars in Scandinavia and in Taiwan, where its subsidiary, "Thai Swedish Assembly", had been assembling them since 1986. But Renault and Volvo decided to cooperate in manufacturing and sales within Far-Eastern Asia: the Volvo network would distribute Renault cars in Australia - where Renault was absent, in Malaysia, where the market is protected, and in Japan. In Great Britain the two groups coordinated their logistic operations and in France, Germany and Switzerland, their sales networks began cooperation, for example over spares distribution.

In conclusion, we can underline the main features of this alliance in the words of one of its architects, R.H. Levy, Chairman and Chief Executive of Renault: "The engagement of Renault to Volvo is really an alliance between equal partners, choosing to put together all their assets, under a common, bi-national and efficient management. One essential aspect of this alliance is the will to create a common corporate spirit: exchange of executives, annual meeting of 200 managers, writing of an 'Alliance charter'". Levy hoped that "this alliance, which has to be managed as a sole enterprise, will know how to exploit all its potentialities, will keep its reaction capability, and remain open to all new ideas, and, why not, to other similar alliances...." (Levy, in the Foreword to Collins and Doorley, 1992). Indeed, from 1991 to 1993, Renault and Volvo implemented a merger process.

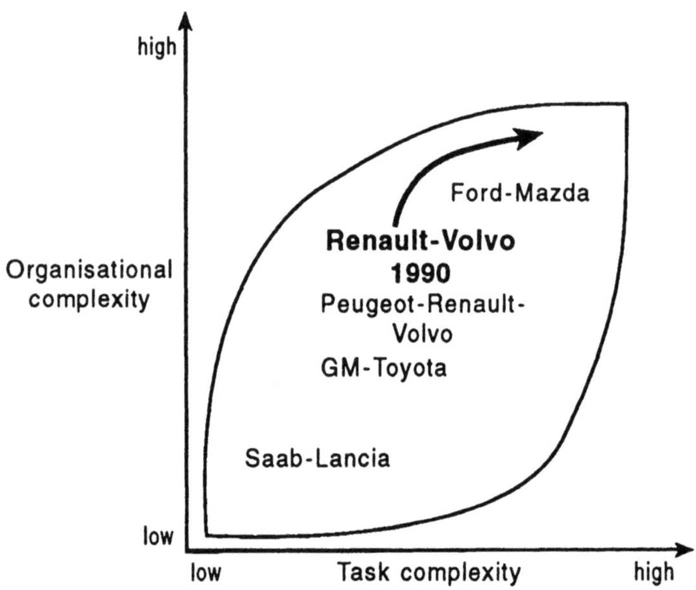

Fig. 7.4 The Renault-Volvo alliance's complexity (in terms of Killing's model).

If we follow the Killing (1988) model for analysing the aims and the organizational complexities of industrial alliances, that between Renault and Volvo would appear, at the end of 1993, as one of the most complex, somewhat similar to the Ford-Mazda one (Fig.7.4). The Renault-Volvo alliance, which was deepened between 1990 and 1993, also appeared to verify the Killing argument about progressive alliance complexity (op cit, 55): "Firms wishing to create an alliance to undertake a complex task should first enter a simpler alliance with their chosen partner, in order that a degree of mutual trust may be established, prior to the formation of the more complex alliance". The continuation of the story suggests, however, that this intensification of the alliance was brought about too quickly.

7.6.2 September 1993: the move towards a merger

The aim of the 1990 alliance was to create a single group, with joint R&D, planning and, manufacturing but with two separate national companies and brands. For three years the executives of the two groups said that the alliance would be developed without merging the companies. As if to emphasise the point, in 1991 Volvo entered into a separate industrial agreement with Mitsubishi jointly to assemble in the Netherlands.

But at the same time there were increasing rumours of merger, reflecting some of the key characteristics of the two groups. Indeed AB Volvo is a diversified group, and the growing losses of its automobile subsidiaries could be avoided by selling them to Renault. Volvo stood to lose $270 million on automobile and truck operations on 1992 sales of $10.3 billion. This was the result of falling sales in its three major markets, the USA, the UK and Sweden (Volvo world-wide unit sales were of 420,000 in 1986, and about 309,000 in 1992). In the automobile sector, in order to cut over-capacity, Volvo reduced its workforce by 1,630 and closed its two modern assembly plants of Kalmar and Uddevalla (Business Week, 23 november 1992; see also Malmberg, this volume).

The French government wanted to privatise Renault, and a merger with Volvo appeared as a necessary step in this. Indeed this merger could be a way, for the French group, to accelerate the process of growth and cost reduction, which had been started with the Volvo alliance. In 1993 that strategic acceleration seemed even more necessary, as the European market was collapsing and the competition rapidly increasing.

Indeed on 6 September 1993, Renault and Volvo announced an agreement to merge from 1 January 1994. The project was to create a new group Renault-Volvo by merging all the car and truck activities. The Renault-Volvo group would be registered in France and majority controlled by the French state, but Volvo would have an important minority stake, corresponding to genuinely shared management control. Louis Schweitzer, Chairman and Chief Executive of Renault would be the chairman of Renault-Volvo RVA, and Pehr Gyllenhammar, his counterpart at Volvo, would be the chairman of the supervisory board of Renault-Volvo RVA. One industrial holding, Renault-Volvo, would be created, in which the French state would own, directly or indirectly, 65 per cent of capital and AB Volvo 35 per cent.

This holding company would control all the former industrial and service activities of Renault and Volvo in cars, trucks and buses. But two different brands, product ranges and sales networks would be preserved. This merger would create a European group, ranking sixth in the world car industry, and controlling 12.1 per cent of the European car market. The new group would have tightly to coordinate all operations; that is to say, to improve on what had already been started in three years of alliance. Important restructuring operations could be speeded up, mainly in truck manufacturing, which was facing overcapacity. It was hoped that managerial efficiency would be increased through this merger, because "managing and co-ordinating the two firms' activities through a stifling bureaucracy of Franco-Swedish committees was too cumbersome" (The Economist, 11 September, 1993). The merger would allow streamlining of decision making processes.

The French government was expecting to privatise the Renault-Volvo group rapidly, by selling its capital stake during the first part of 1995. Renault-Volvo could then soon enter into another financial alliance with the French Matra group, which by then would already be manufacturing the Renault Espace.

7.6.3 December 1993: the merger failure

Following the September announcement about the Renault-Volvo merger project, several Swedish shareholders started an anti-merger campaign. They developed two main fears and arguments. Firstly they were worried that the privatisation of Renault was uncertain and might be delayed for many years. Secondly, they pointed out that the French government would hold a "golden share" in Renault, and be able to block, for social or political reasons, major strategic changes in the merged company.

So the opposition grew within the Swedish shareholders, in spite of French concessions: the French government finally decided that the privatisation of Renault would be in the second quarter of 1994, just after that of Elf-Aquitaine. It also accepted that, after an eight year period, the Volvo voting rights within Renault-Volvo would be able to rise from 20 per cent to 35 per cent. However the anti-merger campaign received much support from Swedish public opinion, based on nationalistic fears, and in business quarters, because of the fear that a state-owned company would be unable to effect necessary restructuring operations: "In France, while many state owned companies, including Renault, have made major progress, they badly need to take the additional restructuring steps that are possible only after complete privatisation. Volvo should bear that in mind before accepting a wedding ring and then finding itself in an unworkable arrangement"(Business Week, 22 November 1993). The outcome of the vote in the Volvo's shareholders assembly, fixed for 9 December, became more and more uncertain.

So, on 2 December 1993, Pehr Gyllenhammar, Chairman and Chief Executive of the Volvo group, tendered his resignation, and the board of the Volvo group decided to cancel the vote about the merger. The initial comments made over the following days by the Renault and Volvo managers showed that the merger was not postponed but cancelled. Indeed, within a few months, several decisions were taken quickly to undo the major "pre-merger" strategic cooperation developed during the three previous years. In January 1994 the joint purchasing committee, created in July 1993, was dissolved, and the common purchasing policy ceased to operate. The joint quality committee was also abolished. The joint development of a common platform was suspended. Finally on 17 February 1994 Renault and Volvo agreed a mechanism for ending their cross capital participations, through the future privatisation of Renault.

The only forms of cooperation still intact in March 1994, and which could continue, were the industrial ones: exchanges of engines, gearboxes and car bodies, which bring cost reductions but allow independence. Of course the two groups will not now benefit from joint R&D, joint purchasing, and all the joint restructuring operations which could have been implemented in a unified company: they were planning to save FF 30 to 40 billion via scale economies as a result of the merger.

One of the main reasons for the failure seems to have been the development of doubts and opposition to the merger inside the Volvo management, partly due to cultural differences. If "trust is the essence of cooperation" then "the importance of joint ventures, alliances and cooperative forms of international business is that they encompass elements of cooperative behaviour. The essence of cooperation is trust, and trust can best be analysed within a cultural context"(Buckley, 1991, 18). The "pre-merger" period may have been too short to allow enough reciprocal exchange of knowledge and mutual learning about different organizational cultures.

Of course this unbalanced merger, which looked like a take over of Volvo by Renault, could easily generate strong opposition from Volvo managers. The uncertainty about the future stable shareholders of Renault after its privatisation also created important doubts which could have been avoided if Renault had been privatised before launching the merger process.

Does this failure of the Renault-Volvo merger invalidate the thesis of this chapter: the need for close cooperations and mergers between European manufacturers? Not at all. It only underlines the important difficulties that confront intra-European mergers, due to nationalistic reactions. Indeed the alternative of acquisitions by non-European groups, which was common in the 1970s and the 1980s, could continue. Analysts are predicting that the Volvo car division could not survive without another alliance, for instance with Mitsubishi, with whom it is already cooperating for designing and manufacturing a new middle range car in the Netherlands. Renault could also cooperate with other groups, such as Fiat.

7.7 Privatisation and After?

The next important event for the development of Renault will be its privatisation. The French government finally decided in March 1994 that it will occur in 1995. This privatisation will give Renault a greater freedom for making strategic choices, and in particular international alliances. These alliances appear necessary, mainly in the loss-making trucks sector, but also in cars. Who could be the partners?

7.7.1 Towards a Renault-Fiat deal?

On 29 March 1993 Business Week announced that Renault and Fiat had secretly drafted a blueprint to create an alliance between them, through a $5.3 billion Fiat share offering reserved for Renault. That would give the French carmaker 40 per cent of the Italian group. At a later stage, the Agnellis themselves could sell some or all of their Fiat stake to Renault. This news, which was based on sources close to Fiat Chairman Gianni Agnelli, was promptly called "fantasy" by Renault's Chairman Schweitzer, and Agnelli said that no big deal was being contemplated "for the moment". In 1994, after the failure of the Renault-Volvo merger, rumours of talks between Renault and Fiat reappeared.

If Renault and Fiat enter such a merger, it would create a continental giant, building almost one out of every four cars manufactured in Europe. It would become the major European group, with more than 19 per cent of the European market, and strong positions in France, Italy, Spain, Sweden, the former Yugoslavia, Turkey, Poland and the new Russian independent states. And Renault-Fiat would become the fourth largest car producer world-wide, with more than 4.3 million units a year, just less than General Motors, Toyota and Ford, and ahead of Nissan and VAG (Fig. 7.5).

Several facts explain why the two groups may wish to enter such a deal. Among all the major European producers, Fiat is the most dependent on its domestic market. So it is the most vulnerable to the removal of national and EC market protections: in 1987 the Japanese groups held only 1 per cent of the Italian market, 15 per cent of the German market, 11 per cent of the British market: see Hudson, this volume; also Conti and Enrietti, this volume). After selling its stake in the Spanish company Seat and rationalising its Latin

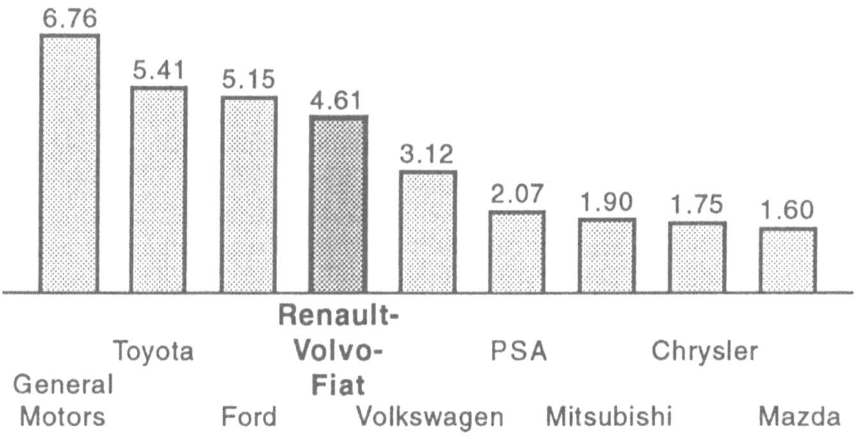

Fig. 7.5 The world's top 9 car manufacturers in 1991 (million units per annum).

American operations, it is little internationalised: for instance in 1990 its world production was of 1,862,000 units in Italy, and 259,000 units abroad (see Dicken, 1992a, 25). Its major international investments are in eastern Europe, mainly in Poland and the former Soviet Union. The merger could give Fiat a more international dimension. Another Fiat weakness is its overdependence on the smaller-car market segments where competition is most fierce. Renault could facilitate a move towards upper market segments. Furthermore, Fiat is ripe for change, because Chairman Agnelli, patriarch of the family that holds 40 per cent of the company, and Managing Director Cesare Romiti are both due to retire in 1994. At the same time Fiat is under financial pressure to fund a long-term programme to retool factories and launch 16 new models over the next few years. In September 1993 Fiat, facing growing losses and indebtedness, doubled its capital stake, found new secure shareholders (Deutsche Bank, Alcatel-Alsthom and Generali), and sold its store chain Rinascente.

On the other hand, for Renault, which failed in its alliance with Volvo, acquiring 40 per cent of Fiat's capital could be the basis for developing a new alliance in Europe. This could help it to grow more quickly, and enlarge its international operations. Indeed the market crisis is severe for Renault: in 1993 its total production volume was reduced by 250,000 units, equivalent to one assembly plant.

But both French and Italian politics could limit or forbid the deal, mostly because several Fiat executives have been implicated in the anticorruption investigations which are reshaping the political and economic face of Italy. The European Community anti-trust officials could also oppose of such an agreement. Managers of the two groups may also hesitate to cooperate, because Renault and Fiat are two similar generalists, which are competitors across the product range and because they remember the failed Renault-Volvo merger.

7.7.2 Towards new international alliances within Europe

Whatever happens between Renault and Fiat (full merger, alliance or nothing), that possible project reveals the trend for the coming years. As the implementation of the Single European Market is removing national and EC protections in automobile markets (see Sadler, 1992, and this volume), Japanese competition will increase. Renault and the other European manufacturers, which are very dependent on their home market or/and on the European market, will have to find ways to combine their strengths: specialised agreements as in the 1980s; balanced broad alliances, like Renault-Volvo in 1990-93; balanced merger; or simply take over.

The potential mergers, such as Renault and Fiat or Peugeot and Volkswagen, or possible acquisitions such as Mercedes-Benz and Fiat (Business Week, 7 March, 1994), would create world-sized European companies, with total sales similar to the world leaders Ford, General Motors and Toyota. But they will remain only really internationalized within Europe. To succeed in a more and more "global industry" they would have to become truly global competitors, by setting up facilities and gaining market shares within North America and Japan.

At the end of year 1993, the fears of the Swedish Volvo shareholders, fighting against the Renault-Volvo merger, emphasized the strong obstacles standing in the way of merging "national champions". Such opposition and failures do not mean that such mergers are not the future of the European automobile industry (Business Week, 7 March 1994). Indeed in January 1994 the German specialist BMW purchased from British Aerospace its 80 per cent capital stake in Rover, despite the 20 per cent participation of Honda in Rover (see Hudson, this volume). This take-over doubled BMW's share of the European market (to 6.4 per cent), and created a new strong European competitor. At the same time Ford and Mazda were deepening their cooperation which is moving from partial agreements to a more strategic alliance. All the Japanese groups, which are facing important losses, are also developing cooperation agreements with one another, inside Japan, in order to reduce costs.

Chapter 8
The Restructuring of the Swedish Automobile Production System

Anders Malmberg
Department of Social and Economic Geography, Uppsala University, Sweden

8.1 Introduction

For most of the 1980s, the Swedish car producers Volvo and Saab were remarkably successful in marketing relatively exclusive cars in the USA, as well as in Sweden and on certain European markets. From 1987 onwards, much of this changed. Saab's overseas sales more or less collapsed and, as a consequence, 50 per cent of Saab Automobile was sold to General Motors in 1990. Since then, Saab has experienced different forms of rationalization, vertical disintegration and spatial centralization. A number of functions and operations previously handled within the company have been externalized, while others have been spatially integrated in Trollhättan, the location of the company's head office and main assembly plant. During 1992, Volvo was also forced to make radical cutbacks in production.

The European car manufacturing industry is to a large extent still made up of several basically national production systems. Despite the process of internationalisation, the national market is still the most important to many European producers of cars, partly as a result of protectionist policies at national and EC levels (see Sadler, this volume). But it is anticipated that the spatial division of labour in Europe will undergo substantial change as these structures are confronted with the post-1992 internal market. The internationalisation of the 'national champions' in different European countries and the seemingly inexorable inflow of Japanese cars, through increasing imports or the establishment of new 'transplants', will imply more intense competition in the saturating European car market. The pressure to increase productivity will rise, accelerating the shake out of the European car production system.

Has there been such a thing as a Swedish, nationally defined, car production system? Does it still exist or is it disintegrating, in the process of European integration and intensified global competition? Will Swedish car producers survive the 1990s, and will cars still be produced in Sweden in, say, ten years time? These are the kind of questions that guide the presentation in this chapter. Recent and contemporary processes of restructuring in the Swedish car industry are described, with particular focus on the changing relations between car producers and component suppliers, and on the regional impacts of the process.

8.2 The Swedish Automobile Production System - Some General Characteristics

The Volvo Company was founded in Göteborg in 1926 as a spin-off from SKF, the producer of ball-bearings. During the 1980s, production of private cars dominated sales as well as profitability, but this has not always been the case. Even though it was the vision of Swedish production of private cars that guided the founders of Volvo, for the first 30 years it was production of trucks and buses that kept Volvo in business. Only from 1950 onwards did private cars become the company's leading product. The period 1950-70 was a phase of major growth of the Volvo company (Ellegård, 1983), accompanied by rapid growth in productivity as well as in production volume and employment.

Saab (Svenska Aeroplan Aktiebolaget; 'Swedish Aeroplane Ltd') was formed in 1937 to produce aircraft for the Swedish Air Force, and took up car production after World War Two. In 1969, Saab-Scania was formed, as Saab merged with one of the two large Swedish producers of trucks and buses, Scania-Vabis. From an early stage, Saab acquired an image as a producer of technically sophisticated and original, though somewhat unusual, car models. With the introduction of the first private car with a turbo engine in 1978, Saab started to enter the upper middle and luxury segment of the international car market.

Sweden accounts for only one per cent of global car production. In 1987, Volvo ranked 18th and Saab 25th on the list of the world's largest producers of private cars. Thus both companies are small in this industry of global mass production. Saab was the second smallest of all independent car producers outside the Eastern Bloc. Nevertheless, it is truly extraordinary and unique that Sweden, a country with only 8.5 million inhabitants, has been, until very recently, the home country of two independent car producers.

If Swedish car production is modest in size globally, it is the more important in the national context. Volvo is the second and Saab-Scania the fourth largest industrial group in Sweden (ABB is the largest, Electrolux the third). During the last few decades, the automotive industry has shown many characteristics typical of a propulsive industry, being in a Perrouxian sense a central focus of a growth pole (Perroux, 1955). The sector (including cars, trucks and buses) accounts for more than 10 per cent of Swedish manufacturing industry, in value terms as well as employment, and the induced growth effects in other sectors (steel, engineering products, rubber, plastics, textiles etc.) are widespread. The value of car exports is almost SEK 50 billion, some 15 per cent of total Swedish exports. Eight out of ten Swedish cars are sold outside Sweden, primarily in North America (42 per cent) and Europe (35 per cent). In 1987, 432,000 cars were produced in Sweden. Since this 'all time high', production volume has decreased considerably (Ministry of Industry, 1990).

To apply the conceptual framework developed by Porter (1990), the 'diamond' around the Swedish automobile cluster began to take shape after World War Two. By this time, important supporting and related industries developed, such as bearings, gear boxes, high quality steel and engines. Rapidly increasing prosperity in Sweden during this period made domestic demand more sophisticated. Market segmentation oriented Swedish firms towards safe and reliable cars, trucks and buses (Sölvell et al., 1991). Not least in the case of safety aspects, Swedish consumers have anticipated trends in international demand. For example, Volvo was the first car producer to introduce the lap-diagonal seat belt in 1958. The firm that developed this product, Autoliv, today a subsidiary of Electrolux, is one of Sweden's more successful car component producers (seat belts, airbags etc.) with more than 3,000 employees and operations in ten countries. A more recent example is Volvo's introduction of a new side impact protection system (SIPS) in 1991.

The establishment of two independent car and truck manufacturers, belonging to two different groups of owners, enhanced dynamic domestic rivalry. Saab-Scania has always been controlled by the Wallenberg family, while the ownership of Volvo has been, and still is, more diversified. More than most other Swedish clusters, the car industry with its active rivalry has retained an environment for sustained innovation and competitiveness.

8.3 Volvo and Saab: Corporate Structures and Strategies

In many respects, Volvo and Saab-Scania have similar corporate structures. Both companies operate in the same product areas and develop, manufacture and market an almost identical mix of goods: private cars for the upper market segment, heavy trucks and buses, as well as engines for marine and industrial applications. Both companies also operate in the aircraft industry. All major product lines expanded during the 1980s, and both companies have kept their production base for cars and engines in Sweden, with some 30 per cent of the assembly located abroad: Volvo in Belgium and Saab in Finland. The production of heavy trucks is more globally dispersed. Apart from the Swedish operations, production plants are located in Europe (Volvo in Belgium, Scania in the Netherlands), Latin America (Brazil) and in the case of Volvo also in the USA.

There are, however, differences as well as similarities between the two companies. Volvo has been more active in its internationalisation with a number of foreign acquisitions, and the company's turnover and employment is almost twice that of Saab-Scania. Even though most structural characteristics and strategic choices of Volvo and Saab are similar to those of other European car manufacturers, there are some distinguishing features.

8.3.1 Diversification

Volvo and Saab-Scania are more diversified than most other car producers[1]. Figure 8.1 shows how the 1991 sales of Volvo and Saab-Scania are distributed among product markets. When trying to delimit the Swedish car production system - no matter whether it is defined as a cluster, a value-chain or a production *filière* - this diversification causes some problems. For both Volvo and Saab, the production of trucks and buses is much more important than for most other car producers. This is particularly true for Saab-Scania. During most periods production of heavy vehicles has been at the centre of this group's operations, technologically as well as economically. Volvo is the second and Saab-Scania the fifth of the world's largest producers of heavy trucks. Together they account for some 20 per cent of global production. The company that in this respect is most similar to the Swedes is Mercedes-Benz - the world's largest producer of heavy trucks and buses, and at the same time one of the

[1] In addition, the 1970s and early 1980s saw the entry of Volvo into new sectors, most notably the food industry, pharmaceuticals and trade (oil and fruits). This diversification was primarily aimed at protecting Volvo against the cyclical downturn in motor vehicle sales. In 1989, however, Volvo sold off its trade operation and later withdrew from pharmaceuticals and the food industry as well. In the latter cases, Volvo exchanged its shares of Pharmacia (drugs) and Provendor (food) for a 42 per cent stake in Procordia, the former Swedish state holding company. This was because diversification had turned out to be economically relatively unsuccessful. Scandinavian Trading and Provendor accounted for 22 per cent of Volvo's sales, but only 2 per cent of the profits of the group. It is also generally believed that Volvo needed to concentrate its resources before it could contemplate entering into partnership with Renault (see below).

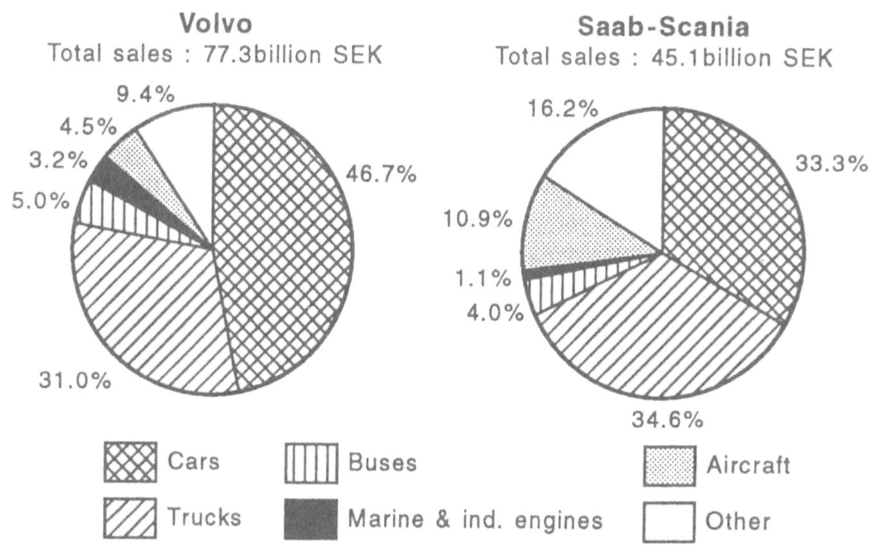

Fig. 8.1 The 1991 sales of Volvo and Saab-Scania (including Saab Automobile) by product areas. (Source : Company Annual Reports).

large producers of private cars (Berggren, 1990). Even for Mercedes-Benz, however, the production of private cars has been the profit centre that has enabled the company to subsidize other operations.

Economically and technologically, as well as in relation to component suppliers and various institutions, the production of private cars is intertwined with the production of trucks and buses, and to a more limited degree with the aircraft operations of both Volvo and Saab. Consequently, the restructuring of Volvo's and Saab's car production, which is the main focus of concern here, cannot be properly understood in isolation from what is going on in the other branches of activity within the companies.

8.3.2 Internationalisation

With respect to cars, the combination of a limited domestic market and strong specialization has led to an internationalisation process, which has taken various forms. First of all, import penetration in the Swedish car market has always been high, compared with other car producing countries. As early as 1970, 50 per cent of the cars sold in Sweden were imported. The corresponding contemporary level of import penetration was approximately 20 per cent in Germany, France and the USA. In 1986, 30 per cent of cars sold in Germany were imported, and the share in the USA was somewhat lower. In the UK, more than 50 per cent of the cars were imported, as domestic car production more or less collapsed during the 1970s (see Hudson, this volume). In Sweden the proportion of imported cars reached 70 per cent in 1986, despite the strong growth of Swedish car production during the preceding period (see Dicken et al., this volume).

Swedish car exports increased as well. The initial strategy of Volvo, in the 1930s, was to produce ordinary cars, suitable for the Swedish climate[2]. As car sales at that time were very much concentrated in spring and summer, Volvo started to export winter production to countries where the seasonal variations were different from those in Sweden. Another explicit strategy was to export only to countries without domestic car production. These were two reasons why Latin America, and in particular the Argentine, became an early important market for Volvo (Kinch, 1992). This original strategy was abandoned in the 1950s when exports to USA began. By 1970 some 65 per cent of Swedish car production was exported. By 1986 more than 80 per cent of the cars produced by Volvo and Saab were sold outside Sweden. The corresponding level of exports was 60 per cent in Germany and less than 50 per cent in Japan (Berggren, 1990).

In 1963, Volvo started to assemble cars outside Sweden, as its previous marketing company in Canada also took up production (in Halifax). Assembling was also located in Ghent, Belgium, today Volvo's largest assembly plant, along with Torslandaverken in Göteborg. Volvo also started - in the regime of another company - to assemble cars in Malaysia. Another dimension of internationalisation concerns component supply. While car producers in Germany and Japan are surrounded by a strong and R & D intensive component industry, Swedish car producers have always relied heavily upon imported components, especially technologically advanced components. With a few exceptions, Swedish component suppliers are predominantly producing relatively simple parts, rather than advanced systems. This gives the Swedish car component industry a somewhat fragile structure.

---

[2] This means that the cars must function in severe cold and on slippery roads. The Swedish climate has recently contributed to an interesting development. The small towns of Arvidsjaur and Arjeplog in northernmost Sweden have become an international centre for winter testing of cars and components (such as ABS and anti-spin systems). In total some 500 kilometres of special winter roads are prepared on frozen lakes, and most European and a few Japanese producers of cars (such as Mercedes-Benz, BMW, Opel, Audi, Fiat, Volvo, Toyota) and car components (for example, Bosch, Bendix, Wabco Westinghouse, Michelin, Bridgestone) - in total more than 40 companies - operate test camps in the area during the winter months. This has meant considerable investment in infrastructure as well as an economic upswing for several local service firms. Some 200 people in this remote part of Sweden make their living through this activity (County of Norrbotten, 1989).

Intensified co-operation, in various forms, with car producers outside Sweden is yet another significant aspect of the internationalisation process. Volvo acquired the Dutch company DAF in 1975, which subsequently became NedCar, owned jointly with the Dutch state and the Japanese company Mitsubishi. During the last decade it has also been involved in various alliances, with, for instance, Peugeot, Renault, British Leyland and General Motors. In the early 1990s, Volvo and Renault agreed on co-operation and co-ownership, and an integration process was started that, for a time, looked as if it might well end up as a full merger between the two companies. The ownership of Saab's car production since 1990 has been divided equally between Saab-Scania and General Motors (see below).

8.3.3 Production organisation and location

The main strategy during Volvo's early years was to buy components from independent and competing firms (Ellegård, 1983). Volvo, located in Göteborg, intended to remain an assembler of purchased components - this was one of the main ideas behind the slogan 'the Volvo way of producing a car'. This strategy was, however, already partly abandoned in 1930 when Volvo bought the engine producer, Pentaverken, in Skövde, some 150 kilometres northeast of Göteborg. Since then, Volvo has taken over subcontracting firms that have proved to be difficult to alter, have implied large costs or have been unable to afford the investment programmes considered necessary. In 1935, Volvo was separated from SKF and became an independent company. In the 1960s, Volvo's subsidiaries within the sphere of car production were merged with the parent company, in order to strengthen coordination and control over production.

In 1958, Volvo established its first branch plant, when production of brake drums was located in Floby, 100 kilometres northeast of Göteborg. Especially during the 1970s, Volvo established a number of new plants in Sweden (Bengtsfors, Lindesberg, Kalmar, Vara, Tanumshede, Färjelanda) - all except Lindesberg (mid-Sweden) and Kalmar (southeast Sweden) located within two hours' drive (that is, 150 kilometres) of Göteborg, and all of them in rather small, previously not heavily industrialized, towns. In 1991, Volvo employed 43,600 persons in Sweden. Almost half of them were in Göteborg, but Volvo's production system involves many places in Sweden. Volvo is present as an employer in every fourth Swedish municipality (commune), and more extensive operations are found in some twenty communes.

Volvo's annual production (excluding NedCar) of more than 300,000 cars (two thirds of which are assembled in Sweden), more than 50,000 trucks and 4,000 buses and coaches makes up, evidently, an extensive and complex production system. From a quick glance at the map (Figure 8.2) we can conclude that production is spread all over southern Sweden. The most

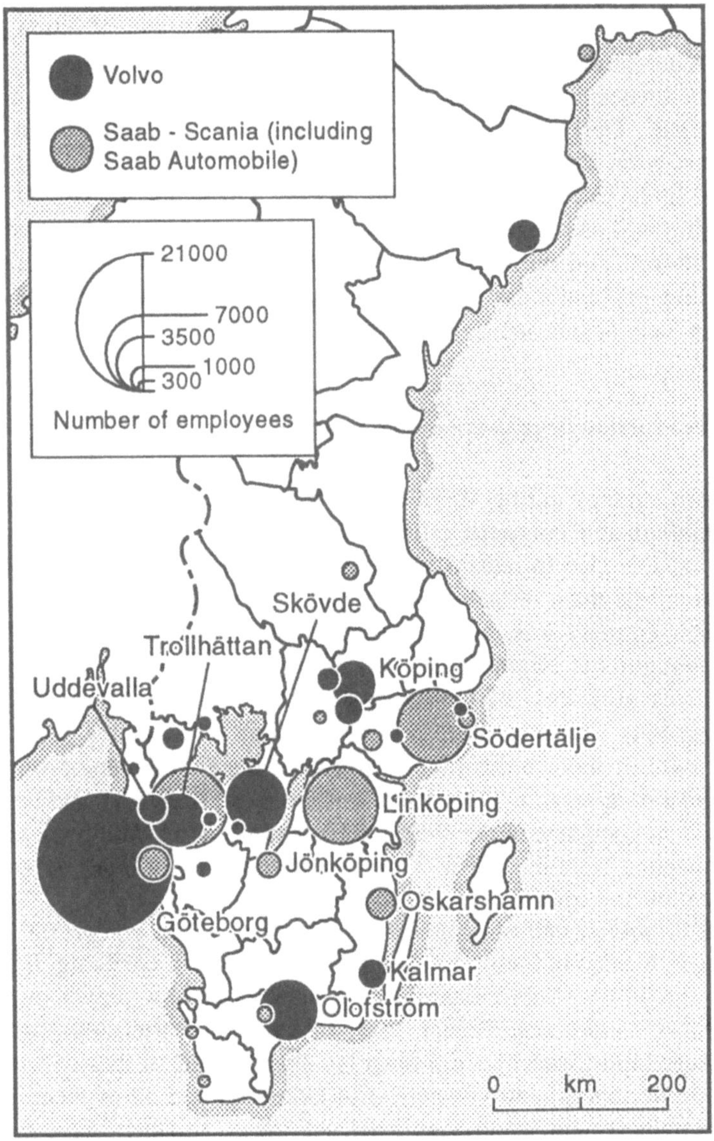

Fig. 8.2 Location of the Swedish operations of Volvo and Saab-Scania (including Saab Automobile) in 1991. (Source: Company Annual Reports).

striking feature, though, is the concentration in south-western Sweden and the Göteborg area. Plants involved in the production of private cars tend to be especially clustered in this region.

Saab-Scania's production of trucks is concentrated to Södertälje, close to Stockholm, and the aircraft operation is located in Linköping. Saab

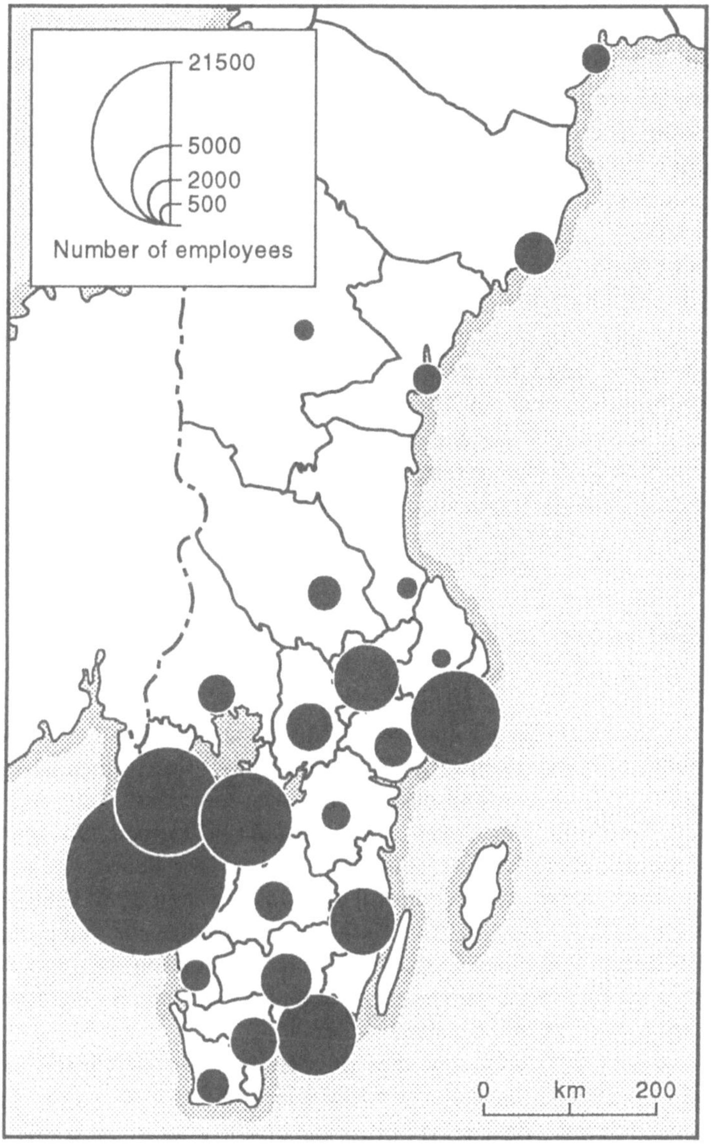

Fig. 8.3 Employment in the Swedish automobile industry (ISIC 3843) by county in 1991. (Source: Statistics Sweden).

Automobile's car production system is less extensive than Volvo's. Through rationalization, employment decreased from almost 17,000 to less than 13,000 between 1989 and 1991. The main plant in Trollhättan employs 7,700 persons, and other major plants are located in Göteborg (gear boxes) and Södertälje (engines). Since the closure of the new assembly plant in Malmö in

1991, the only remaining assembly plant outside Trollhättan is in Nystad, Finland, where the 900 Model for the Finnish market is assembled. GM has recently located the assembly of the Opel/Vauxhall Calibra at Nystad, to compensate for the relocation of most of the assembly of the Saab 900 to Trollhättan. The overall regional distribution of employment in the Swedish automobile industry is shown in Figure 8.3.

The Swedish car production system leaves a large number of towns and municipalities, often small in size, heavily dependent on the commercial success of Volvo and Saab. The latent risks of this dependency do not cause severe problems during periods of steady growth. At present, however, the system is undergoing substantial changes. This is partly the result of a general re-organisation towards a more 'lean' production throughout the global auto industry (Womack, et al., 1990), but mostly because of Swedish car production entering a severe crisis, at the very end of the 1980s. The remainder of this chapter will describe the development of the Swedish car industry prior to the present crisis, and the implications following from efforts to resolve it.

8.4 1970 to 1987 : From Crisis to Success

In the 1970s, small car producers like Volvo and Saab, and more generally independent car production in small countries, were often regarded as relicts of the past. This was a bad decade for both companies. The car models marketed were old (introduced in the 1960s) and exports, not least to the USA, declined. Rapidly increasing wage costs in Sweden in the mid-1970s made the crisis acute. Production fell and was lower in 1980 (250,000 cars) than in 1970 (280,000 cars). The Wallenberg family at one stage regarded a merger between Volvo and Saab as the only solution, but this proposal was dropped in the face of strong resistance from executives in both companies, as well as from many of Volvo's share holders.

In retrospect, it is apparent that the effects of crisis turned out to be positively dynamic. Both Volvo and Saab increased their R & D investments. Existing models were developed further and efforts to create new models were intensified. Saab's 99 Model was upgraded to the 900 Model. Due to the technological abilities within Scania, the Saab 900 became the first private car to be equipped with a turbo engine as standard. For a few years, Saab held a world leading position in this technology.

The onset of the 1980s brought a dramatic and positive change. During the first half of this decade, both Volvo and Saab were among the world's most rapidly growing and profitable car producers. Between 1980 and 1986 the number of cars produced by Volvo and Saab increased by some 60 per cent. In value terms growth was even more rapid. This was a higher growth rate than that achieved by German producers competing in (more or less) the same market segment. During the 1980s, seen as a whole, the combined turnover of Volvo and Saab-Scania tripled, and employment increased by 20 per cent. In 1986, both Volvo and Saab decided on major investments in new assembly plants, located in Uddevalla and Malmö, respectively.

There were a number of reasons for this success. One is that both Swedish car producers had managed to shift from a previous concentration on production of ordinary cars for average income households to the 'upper middle segment' of the market and more luxurious cars for people with higher incomes. The USA definition of this segment in 1986 was cars priced more than $15,000 US. This turned out to be a successful strategic choice, since this segment of the market proved to be less affected by decline following the oil crises, as well as, at least temporarily, by the increasing competition from the Japanese volume producers. On the domestic Swedish market, though, both Volvo and Saab predominantly sell their cheaper models, and these cars are still regarded as ordinary.

The success of the Swedish car industry in foreign markets during the 1980s was also reinforced by macroeconomic and institutional factors. The industry was strongly supported by a series of devaluations of the Swedish currency between 1980-1982. A more long term influence was the 'solidaristic wage policy', an important feature of the 'Swedish model' stemming from the early 1960s. The original aim of this policy was 'equal pay for equal work', but it turned out to be a major force accelerating the restructuring process, since all firms had to pay similar wages, regardless of their profitability. By international standards, wage levels in industries like textiles and clothing became high, while on the other hand wage levels in engineering, and particularly in the car industry, became relatively low. Still, however, relative to western Europe, wage levels in Swedish car production are high (see Hudson, this volume).

8.5 1988-1992: Renewed Crisis and Intensive Restructuring

8.5.1 Saab Automobile

After a number of extremely successful years, Saab saw profits from car production fall drastically from 1987 onwards. In combination with increasing investment costs following decisions in the mid 1980s, this led to an acute crisis in 1989. There were a number of reasons underlying this rapid change. With the introduction in 1984 of its 9000 Model, Saab had gone into competition with high prestige German marques, but obviously had partly misjudged its market position. The 9000 never sold really well outside

Sweden. As well as losing market shares to BMW and Mercedes-Benz, Saab also found itself losing old customers who could not afford these more luxurious cars. In 1989 production was limited to 60,000 cars annually, which was below the break-even point (Berggren, 1990). Saab also found itself competing in the market segment in which Japanese competition during these years was most aggressive.

In addition, Saab had a productivity problem. Traditionally, its models have had a reputation of being the dream cars of construction engineers, rather than of production engineers. The Saab 9000 proved to be more expensive to manufacture and assemble than competing models. The degree of standardization was low, the mix of components was large and diversified, and the assembly of the car was complicated and labour intensive. Also, the production organization itself was inflexible and inefficient. With the opening of the new assembly plant in Malmö, Saab had the capacity to produce 180,000 cars, but could sell only 120,000.

During 1989, it became obvious that Saab would not survive as an independent car producer. It nevertheless came as a great surprise to most observers when, in December of that year, it was announced that a new company, Saab Automobile, would be established, jointly owned by Saab-Scania and General Motors (GM). GM paid some SEK 10 billion for its share in Saab-Scania's private car operation, including the development and manufacture of engines and gear boxes. Under its new American leadership (David J. Hermann from GM Europe was the managing director during the first three years), a number of more or less radical measures were taken in order to restructure production organization, reduce the work force, cut costs and improve productivity, thereby trying to combat the crisis that became even deeper immediately after GM's takeover. During 1990, component plants were closed in Kramfors (in northern Sweden, where electric systems were produced), Kristinehamn (in mid-Sweden, where axles and seats were produced) and Halden (Norway, electric systems), and Saab's seat operation was sold to Lear Seating of the USA.

The most radical decision, though, was to concentrate production at the main plant in Trollhättan. Thus, the brand-new assembly plant in Malmö was closed without having ever been run at full capacity[3]. A decision made in 1989 to build a new engine factory in Karlskrona was abandoned. During the 1980s, both Malmö and Karlskrona were severely hit by the collapse of the

[3] The 95 000 m^2 factory, newly built at a cost of 1.3 billion SEK on the site of a previous shipyard in Malmö, has been empty since 1990. All attempts to find a buyer failed. Towards the end of 1992, Saab started to look for potential tenants and the first contract was signed in November. The first tenant is the 'Malmö Golf Academy', which will use a small corner of the factory for an indoor driving range. There is now hope that others will follow and that parts of the factory will be turned into a sports centre.

shipbuilding industry, and Saab had received considerable financial support from the Swedish government for these new investments. Saab's decision to withdraw from these commitments thus caused some tension in the relation between the company and the government.

Saab's sales and marketing organisation was also centralized in Trollhättan. The head office of its marketing organisation was previously located in Nyköping, on the east coast, south of Stockholm. Marketing in North America, Europe and the Far East is now co-ordinated from Trollhättan, and only the Scandinavian market organisation (including Sweden) was left in Nyköping (Saab Annual Report 1991). Despite the relocation of various operations to Trollhättan, the work force there has decreased, from almost 8,000 to just less than 6,000. According to information from Saab, labour turn-over and absenteeism have gone down radically, while there has been a considerable rise in quality and productivity. The number of hours to assemble a car has fallen from 115 to 60.

Retailing is another area where the merger with GM has had considerable effects. The number of car dealers in Sweden has been reduced, as a single dealer now markets both Saab and Opel in most cities and regions. In 1991, there were in Sweden 95 car dealers selling Opel and 60 selling Saab. After two years of fierce local competition over which companies would become the dealer for both Opel and Saab, there were, early in 1993, in total only 110 dealers left. The losers in this competition, mostly previously Opel dealers, in many cases have gone out of business.

From the point of view of national industrial policy, a major concern with regard to the consequences of GM's acquisition of Saab relates to component supply. Saab Automobile thereby gained access to GM's world wide supplier network ('GM Global Sourcing'), and one explicit aim of the strategy is to reduce purchase costs considerably. The aim is, of course, to become more 'lean', and the measures taken are those that seem to have gained almost global acceptance in the car industry. Thus, the number of suppliers is reduced by shifting to 'sole sourcing' and organising suppliers in different tiers. The first-tier primary suppliers are forced to take on responsibility for product development and quality control. Saab has evaluated the performance of all its suppliers, and many of those that fall short of expectations will be dropped. In 1990, Saab Automobile had some 480 component suppliers, half of which were in Sweden. According to the company's stated goals, this number will be dramatically reduced within a few years, but it is as yet impossible to assess the real impact of this plan. As a matter of fact, up to the end of 1992 the number of primary suppliers increased rather than decreased, partly as a result of externalization of previous 'in-house' component supply.

Despite all these measures of rationalization and increasing productivity, Saab's crisis was far from over. During 1992, Saab Automobile made a loss of another SEK 2.7 billion. The company had thus lost some SEK 10 billion, in total, during the period 1989-1992. Put another way, every Saab car sold during the four year period resulted in - on average - a loss of 25,000SEK. During the same period, the owners, GM and Saab-Scania, provided Saab Automobile with new capital totalling SEK 13 billion. At the end of 1992, a decision was taken by the owners to add another SEK 2-3 billion, explicitly stating, though, that this would be the last such tranche of expenditure. In November 1992, Saab Automobile announced that another 2,000 workers would be laid off.

8.5.2 Volvo

In 1985, Volvo was probably the world's most profitable car producer. More than 100,000 cars were exported to the USA alone. All production capacity was fully utilized, and a decision was made to build a new assembly plant in Uddevalla, some 100 kilometres north of Göteborg. This was the first greenfield investment that Volvo had made since the Kalmar plant was opened in 1974. As with that at Kalmar, the Uddevalla plant has gained considerable national and international attention, since there is no assembly line. Instead, entire cars are assembled by teams of multi-skilled workers[4].

In 1986, profits began to decline, but a reasonable level of profitability was sustained throughout the decade. Contrary to the strategy of Saab, Volvo never entered the 'BMW route'. Instead it retained its traditional image, and continued to produce a safe, high quality, 'up-scale family car' (Berggren, 1990). Volvo also maintained its USA sales during the latter years of the 1980s, and in comparison with Saab, it has a stronger position on European markets, particularly in the UK. Stagnating production volumes, rapidly increasing R & D costs and unfavourable productivity development meant, however, that Volvo has lost money since 1990.

The expansion of Volvo's production of private cars has to a large extent taken place outside Sweden. In 1980, 140,000 cars were produced in Sweden, and 50,000 abroad. In 1988, 180,000 cars were produced in Sweden, and 110,000 abroad, of which 90,000 were produced in Ghent, Belgium. Subsequently, a decision was made further to increase capacity there. Volvo's new 850 Model is, at least initially to be assembled in Ghent.

During the 1980s, Volvo was relatively strong in production and marketing. The main drawback was its limited model range and the slow development of

[4] Volvo's decision to locate an assembly plant in Uddevalla coincided with the closure of the shipyard that had previously constituted the industrial backbone of the area. Volvo therefore received considerable support from the Swedish government for this investment. In addition to direct and indirect capital support, the 'missing link' of the motorway between Göteborg and Uddevalla was built, to ensure rapid and safe transport. This generosity by the state was later repeated when, partly as a matter of justice, Saab received similar support for its new plant in Malmö.

new models. As there is an extended period between the introduction of new models, the company will face severe problems if there are any market failures.

During this decade Volvo shortened the lead times for its subcontractors as the demand for quick and safe delivery of components increased. For many products, a couple of days may intervene between an order being sent and the component entering the assembly line, but efforts are continuously made to reduce this time. A few subcontractors have a permanent on-line connection to Torslandaverken. For example Sunwind, a producer of floors for the luggage boot of Volvo cars, is located at Säve, some 15 kilometres north of Torslandaverken. Sunwind continuously receives computerized information as to which cars are entering the assembly line at Torslandaverken. Within a couple of hours the floor is delivered, to coincide with a particular car reaching the floor-fitting station. It is planned that this kind of 'sequence-delivery', will be practised by several subcontractors, but in the foreseeable future it will be the exception rather than the rule. In general, the effects of new purchasing strategies and JIT delivery should not be exaggerated. First of all, Volvo had already rationalized its component supply system in the 1960s and 1970s, and 'sole sourcing' has subsequently been the general rule. JIT has had little impact on the geography of the supply chain, however. By the end of the 1980s, Volvo could not identify a single case in which conforming with JIT delivery schedules had led to the elimination of a supplier, or forced a relocation to the Göteborg region (Jonsson, 1989).

In 1990, an agreement on co-operation and co-ownership between Volvo and Renault was concluded. Volvo could acquire up to 25 per cent of Renault, and the French company acquired 10 per cent of the Volvo Group and 25 per cent of Volvo's private car operations. In addition, Volvo and Renault decided to buy 45 per cent of each other's truck and bus production. For some time, the popular view was that differences in corporate culture - and the political inertia following from French state ownership of Renault - would prevent any real integration of the two companies. Some considerable integration nonetheless took place (see also Savary, this volume).

It was thus decided that Volvo's plant in Skövde would deliver the engines to the new car that would replace the Renault 21 model. Production volume was scheduled at 100,000 engines in 1993. Volvo's car body plant in Olofstrom became a supplier for Renault's Clio model. More significantly, Volvo and Renault began a process of systematically reviewing their plans for development of future models, and agreed a joint model programme for the next 12 to 14 years. The fact that the production systems of the two companies were being integrated indicated that Volvo and Renault were actually heading towards a merger and, in September 1993, the companies launched such an agreement. After several months of intense public debate in Sweden, however, Volvo withdrew from this agreement in December 1993, when it became clear that the proposal had been rejected by its shareholders. Volvo's chairman, Pehr Gyllenhammar immediately resigned, along with several other members of the board. There will therefore be no merger and at present the future of the strategic alliance between Volvo and Renault - probably soon to be privatised - is very uncertain.

As well as exploring links with Renault, Volvo has been restructuring its operations in other ways. It is attempting to follow the strategy of, for instance, Toyota and Nissan, in strengthening its control over marketing and distribution. Thus the contract with British Lex Service to market Volvo cars in the UK has recently been cancelled, whereafter Volvo will build its own marketing and distribution network.

Like Saab, Volvo has been forced to restructure its production system during the 1990s. Between 1990 and 1992, its work force in Sweden was reduced by almost 20 per cent, and some production plants have been sold. In the autumn of 1992, Volvo decided, in response to a declining market, to reduce production capacity radically by closing the assembly plants in Uddevalla and Kalmar, and to reduce the number of workers in most other plants in Sweden. The closures in Uddevalla and Kalmar are estimated to have reduced Volvo's costs by SEK 350 million. 1,600 workers have lost their jobs as Volvo reduced production capacity by 80,000 cars per year. The remaining assembly plants in Göteborg (Torslandaverken) and Belgium (Ghent) have a capacity of 170,000 and 130,000 cars per year, respectively. In total, another 4,000 jobs will be lost.

The decision to close the plants in Uddevalla and Kalmar was controversial, not only because Uddevalla was Volvo's most modern plant, but also because both plants have been innovative and progressive in developing alternative ways of organizing assembly work - something that attracted national and international interest[5]. Volvo claims that the decision to close down Uddevalla and Kalmar should not be regarded as a failure of the attempt to produce cars without using assembly line methods. According to Volvo's internal evaluation of all its assembly plants, there are no significant differences in labour productivity between Kalmar, Uddevalla and Göteborg. The official explanation is that the plants in Uddevalla and Kalmar were being closed down because they were incomplete - neither of them produced car bodies and neither of them had painting facilities. It is also a matter of how much production capacity Volvo will need in the future. Torslandaverken has twice the combined capacity of Kalmar and Uddevalla - so a closure of Torslandaverken would be problematic for this reason as well. Finally, the closures can also be seen as a typical 'branch plant case'. Both Kalmar and Uddevalla were branch plants, established to increase capacity in times of high demand. As such they were, according to experience, vulnerable and exposed to the risk of being closed first in times of declining demand.

[5] 'Kalmarism' has come to represent the idea of 'car manufacturing with a human face'. Lipietz (1992) goes even further and assigns the label 'Kalmarism' to the economic and institutional system that is more often referred to as 'the Swedish (or Scandinavian) model'.

8.6 Post 1992: Future Prospects for Swedish Automobile Production

At present, neither Volvo nor Saab are able to produce and market private cars profitably, so the crucial question still remains: will there continue to be production of private cars in Sweden? And if the answer to this question is positive: will Saab and Volvo be considered as Swedish companies or will the internationalisation of the production system and the tendency towards intensified integration with foreign car producers cause them to lose their national identities?

8.6.1 Lost national identities?

Car production is a matter of national pride for most countries in which this industry traditionally has been strong. Governments as well as public opinion are often sensitive to cross-border alliances and co-ownership that expose domestic car production to foreign influence, as was clearly illustrated by the Swedish reaction to the proposed merger of Volvo and Renault in 1993. General Motors' takeover of a 50 per cent share of Saab Automobile is an interesting exception to this general trend. Naturally some worries about future Saab models resembling too closely those of Opel were expressed, but the general reaction from Swedish managers, trade unions and the government was surprisingly positive. This to a large extent reflects the fact that Saab indisputably was in a deep crisis, and that the alternatives discussed (during the autumn of 1989 the most widespread belief was that Saab would be sold to Fiat) were generally considered to be worse[6].

Volvo has a much longer record of international co-operation than Saab. Through its assembly plant in Belgium and the co-ownership of Dutch DAF, which recently also led to a direct link to Mitsubishi, Volvo has been inside the EC market since the mid-1970s. But despite the rapid expansion of capacity in Belgium, Volvo still has its most important production base in Sweden. Some 75 per cent of its recent SEK 16 billion investment programme has been directed towards operations in Sweden. In this respect the strategies of Volvo

---

[6] An anecdote captures the local reaction to the Saab-GM deal in Trolhättan. A few months after GM's take over, the American semi-documentary movie 'Roger and me', which gives a very critical view of GM's way of handling labour relations in Flint, Michigan, was to have its opening night in Sweden. The Swedish film distributor assumed that Trolhättan would be the ideal place to launch this film in Sweden, but after having been seen by only 61 persons during the first week, it was closed down. Clearly the Swedes in general, and those in Trolhättan in particular, very much wanted to believe that General Motors would be a good partner for Saab.

and Saab differ from those of most other large Swedish manufacturing companies during the 1980s, which have located most investments in new manufacturing capacity abroad.

As to the location of future Volvo investments in production facilities, there has been a constant process of playing off Swedish alternatives against Ghent. During much of the 1980s, Volvo's main plant in Göteborg suffered from labour shortages, high labour turnover and a high degree of absenteeism, as did Saab in Trollhättan. From the companies' point of view, this situation has improved during the last few years. This is partly an effect of measures taken by the new conservative/liberal Swedish government, but more a result of the previously 'over-heated' Swedish economy sliding into its deepest recession since the 1930s. Unemployment increased at a faster rate in Sweden than in most other west European countries, rising from less than 2 per cent at the end of the 1980s to almost 10 per cent in 1993.

Volvo's new 850 GLT model has been called 'the last Volvo entirely developed in Sweden', which is of course a severe exaggeration since many components are developed abroad. Even so, the basic construction and design is Swedish. Therefore, it is not without irony that this car is, from the outset, being assembled in Belgium. Not least for Swedish component suppliers, the issue of where the cars are produced is crucial. The same point applies to the development of the new Saab models. Saab has envisaged an entirely new model programme during the 1990s. The greater the GM/Opel content of these new models, the less are the chances for Swedish component suppliers to be able to compete successfully for orders.

8.6.2 Prospects for component suppliers

The European car component supply system is in the early phase of an intense restructuring process (see Sadler and Amin, this volume). Most car

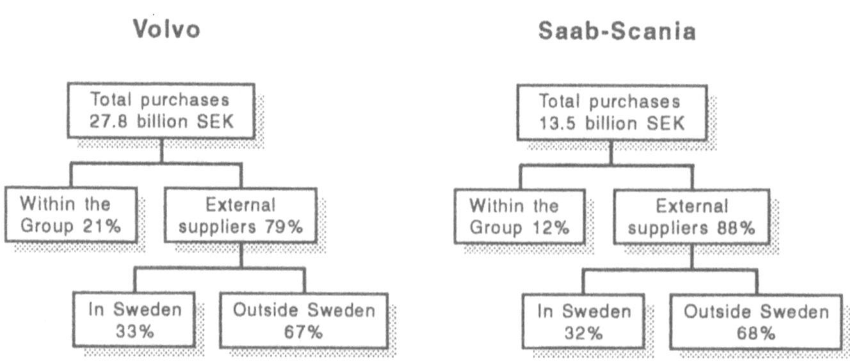

Fig. 8.4 Saab-Scania's and Volvo's component supply. (Source: Ministry of Industry, 1990)

manufacturers are tending to reduce the number of component suppliers. At the same time, there is an ongoing process of externalization. Car producers 'out-source' an increasing share of production and become primarily assemblers, only manufacturing certain key components. The Japanese car producers are generally regarded as the most advanced in this respect.

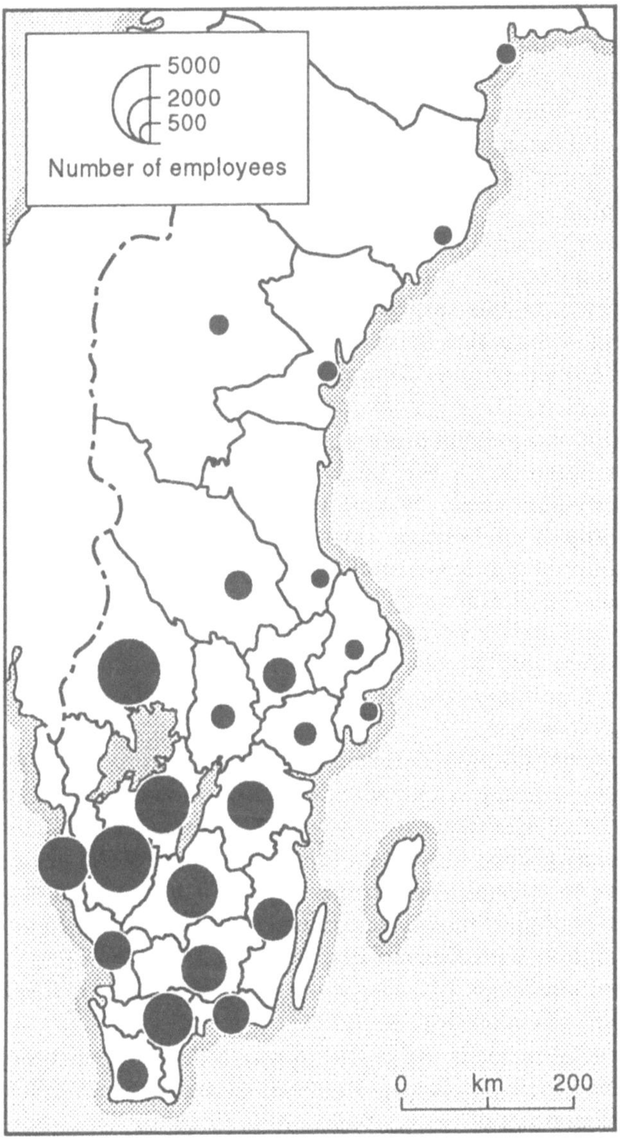

Fig. 8.5 Employment in the Swedish automobile component industry by county in 1989. (Source: Industial Board of Sweden, 1990).

Approximately 80 per cent of the value of a Japanese car is made up by purchased components. The corresponding share in Europe is 50 per cent. Volvo and Saab are, in this respect, rather similar to the Japanese. Purchased goods make up some 70 per cent of the value of the cars produced.

Annually, the Swedish car producers purchase some SEK 40 billion of components from suppliers in Sweden and abroad. Components worth SEK 7.5 billion are produced within Volvo and Saab, and components to the value of SEK 11 billion - corresponding to a direct employment effect of 28,000 jobs - are purchased externally from Swedish component companies. The remaining components (SEK 21.5 billion) are purchased from suppliers outside Sweden. The pattern is broadly similar for Volvo's and Saab-Scania's operations in the automobile sector.

A recent survey of the Swedish car component industry yields some interesting information (Industrial Board of Sweden, 1990). Approximately 350 firms, with some 40,000 employees, make up this industry. Saab is a customer of some 60 per cent of the firms that produce components for Volvo - and vice versa. In this respect, the Swedish car producers are mutually interdependent, and can thus be said to form one production system.

The regional distribution of employment in the car component industry reinforces the picture of a spatial concentration in southern Sweden (Figure 8.5). Göteborg and neighbouring regions account for around one third of employment. There are a few large firms, such as SKF (ball bearings), Swedish Steel (sheet steel), Ovako Steel (castings) and Nobel (plastics), but the vast majority of the firms are small. More than half employ less than 100 persons. A third do not export at all, and less than half export more than 10 per cent of total sales. Only very few of the firms (in all twelve out of 222 in the survey) have a foreign car producer among their five most important customers. From this we can conclude that the Swedish car component companies are unlikely to gain a position as first tier suppliers for the European car producers.

The most crucial factor in determining the future prospects of the Swedish car component industry is undoubtedly the development of the global car market and the competitiveness of Swedish car producers. But a number of other factors can also be identified. The future locational strategies of Volvo and Saab for car assembly, and the extent to which backward disintegration will continue, is of great importance. Equally important is the form of Volvo's future cooperation with Renault and the extent to which this, as well as General Motor's purchase of Saab, will lead to relocation of R&D activities and production system integration.

Another crucial factor is the extent to which Swedish car producers will actually change their purchasing strategies along the lines that have been indicated. The concept of 'lean production' is of course well known to them and both Volvo and Saab have explicitly stated that they will increase the pressure on their component suppliers to take responsibility for product development and at the same time improve quality and reduce prices.

However, it is one thing for a company to express its future intentions in the form of clear-cut strategies, and another actually to implement them. Thus, in 1989 Saab claimed that the number of Swedish car component suppliers would be reduced from 220 to 100 within three years while, in reality, in 1992 the actual number was slightly higher than the original 220. In the long run, however the crucial factor will be the competitiveness of Swedish component suppliers relative to foreign competitors, and the ability of Swedish component suppliers to penetrate foreign markets.

8.7 Concluding Comments

As has already been pointed out, it is truly extraordinary that a small country like Sweden has been home to two independent producers of cars. The Swedish car production system was, of course, never self-contained in the sense that it could develop in isolation from the world market and technological developments originating in other parts of the world. Nevertheless, both Volvo and Saab have developed and produced cars that have been internationally competitive. Seen in the perspective of Porter (1990), Sweden has so far offered a rich environment for the automotive cluster. There has been internal rivalry, certain related industries have been readily at hand, domestic demand has been sophisticated in certain respects (in part because of the rigours of a harsh climate), and anticipatory in others (notably over safety and environmental regulation) and the high labour costs have meant a selective factor disadvantage that has forced the Swedish car producers to be innovative and technologically advanced in production organization.

Nevertheless, there is no doubt that the economic success of Volvo and Saab has been heavily dependent on macroeconomic factors such as relative wage levels and - most of all - exchange rates. The successive devaluations of the Swedish Crown in the early 1980s were important to the success of Volvo and Saab - not least in the USA market - during much of the 1980s. In November 1992, after several waves of speculations against the Swedish crown[7], it was allowed to float. Within a few weeks it had de facto lost some 20 per cent of its value against the ECU and the USA dollar. This produced a great short term improvement of the competitiveness of Volvo and Saab. It coincided with the introduction of Volvo's 850 model in the USA market, and all of a sudden made the prospects of the company much brighter.

---

[7] In trying to defend the value of the crown, the Bank of Sweden during one period raised the short term interest rate to 500 per cent.

Sustainable competitive advantage cannot, however, be based solely on favourable macroeconomic conditions such as an advantageous exchange rate. A necessary precondition for this is the ability of the Swedish car producers constantly to upgrade their innovative capacity through developing new products and more efficient methods of production organization. Throughout the postwar period, south-western Sweden has been the 'home base' of Volvo's and Saab's car production system, and constituted an industrial milieu that fulfilled this precondition. Whether this will continue in the future cannot be determined with any accuracy. There are two critical issues to take into consideration here. The first is whether Saab and Volvo will have the power to survive as independent car producers. The second is whether the industrial milieu in Sweden will be rich enough to retain their 'home bases' there.

From a Swedish point of view, the most optimistic scenario is that Volvo and Saab will remain as car producers with a considerable degree of autonomy, and that south-western Sweden will remain the home base for their operations. This may, however, coincide with more or less extensive relocations of assembly work and the production of standard components from the region. It would mean a continuation of the current vertical, or hierarchical, form of spatial division of labour - corporate management, R&D and production of key components in the corporate 'heartland', and more and more of the actual manufacturing outside Sweden.

Equally possible is that Volvo and/or Saab will lose their identities as autonomous Swedish car producers and will become strongly integrated with foreign car producers, although this seems less likely in the short-term in Volvo's case, following the divorce before the marriage with Renault. This would mean that the function of south-western Sweden would be reduced to production and assembly of cars developed elsewhere. There is a further third option, which is that Volvo and Saab will survive as independent car producers, but that they gradually move their home bases out of Sweden. This would also mean a reduced role for south-western Sweden in the car production system.

In either of the latter two cases, there will be considerable regional effects in terms of reduced employment in the car production system. It is unlikely that - in the foreseeable future - Swedish production will again exceed 300,000 cars per year. This means that the decline in output (-130,000 cars/year) and the job losses registered during the last few years - 10,000 people or 20 per cent of the number employed in 1987 have been laid off - is irreversible. Already rationalization during the last few years has meant severe problems of unemployment in a number of small towns. If the downturn is still deeper, the industrial base of south-western Sweden will be threatened. The real question then is whether we have reached the bottom, or whether the crisis has just started.

Chapter 9
Multi-purpose Vehicles, a New Opportunity for the Periphery? Lessons from the Ford\VW Project (Portugal)

João Ferrão[1] and Mário Vale[2]
[1]Institute of Social Sciences, University of Lisbon, Lisbon, Portugal
[2]Centre of Geographical Studies, the Lisbon Faculty of Letters, Lisbon, Portugal

9.1 Introduction

This chapter explores the qualitative and quantitative significance of the appearance of a new segment in the automobile market - the multi-purpose vehicle or people carrier - and relates it to current global restructuring in the auto industry. Within this context, the Ford\VW Project in Setúbal, Portugal (named AutoEuropa, with each company holding a 50 per cent share) possesses a two-fold paradigmatic value. On the one hand, it clearly illustrates one type of cooperative strategy between American and European companies in response to Japanese competition in the world's largest market, that of the European Community. On the other hand, it reflects the opportunities and difficulties which a project of this kind can bring for a small, semi-peripheral country like Portugal. Consideration of both these aspects allows an identification of new factors involved in redrawing the map of the European automobile production system.

The first section of the chapter summarises probable trends in the evolution of world and European production of people carriers and assesses the relative importance the Ford\VW project in this growth. Portugal's specific situation will be discussed in the remaining sections. A rapid profile of the motorcar industry in this country will be drawn as a background to understanding many of the problems encountered by Ford\VW in seeking to establish a network of appropriately qualified "domestic" suppliers. The AutoEuropa project is then examined in more detail to clarify the sectoral and regional impacts which it might eventually have.

Throughout, it will be emphasised that the Ford\VW project will serve as the catalyst for a new intra-sectoral geography of production in Portugal, at variance with the traditional patterns of inter-sectoral specialization. It is not yet clear whether this new geography of Portugal is simply a local manifestation of more radical changes within the European territory. Only the future can provide a definitive answer to this question.

9.2 Prospects for the Growth of Multi-purpose Vehicles: The Visible Hand of the Single Market

The domestic markets of the most-highly developed countries are close to saturation point. Replacement sales have therefore grown in importance. Due to growth in family incomes and leisure time, there is a strong trend towards a greater demand for high-quality vehicles. Many people also wish to be in closer touch with nature and a revival of "travelling" has generated a new demand for polyvalent vehicles. It is within this context that Noble (1992) identifies "multi-purpose vehicles"[1] as occupying a crucial new niche in the product market.

A marked growth in the number of multi-purpose vehicles is forecast until the year 2000, with global sales of 2.5 million in 1995 - approximately 6 per cent of total sales. The European market for such vehicles has been estimated at between 600,000 and two million, although the latter figure is unlikely as it would represent about 80 per cent of the world market. By the year 2000, the world market for multi-purpose vehicles could reach about 4 million vehicles - some 10 per cent of the overall sales. This would be the same market share that these vehicles currently enjoy in the USA (Table 9.1).

Multi-purpose vehicles are a product innovation in a mature industry. Forecasts of future growth in demand for them have therefore attracted an increasing number of manufacturers to begin to produce them. In the USA, the market is much more developed than in the EC. The main USA producers are Chrysler (with its Voyager, the first multi-purpose vehicle, launched in 1983 and Dodge Caravan, with annual sales of 366,000) which occupies first place, followed by General Motors (Chevrolet Lumina and Astra, Pontiac TranSport and Oldsmobile Silhouette - totalling 250,000 annual sales) and then Ford (Aerostar - 180,000) (Financial Times, 19 June 1992).

[1] This vehicle is a high-roof, 7-8 seat estate car. It is also referred to as a "minivan", "people carrier" and "monospace".

Table 9.1 Multi-purpose vehicle market estimates

Year	World market (millions) Total	World market (millions) Multi-purpose vehicles	European Market (millions) Total	European Market (millions) Multi-purpose vehicles A	European Market (millions) Multi-purpose vehicles B
1990	35.6	1.6	12.1	0.4	0.5
1995	39.1	2.5	13.3	0.6	2.0
2000	43.0	4.3	14.6	1.5	-

Note: Estimates for the motorcar market have been based on forecasts for European production for the year 2000 (CEC, 1992). This figure was taken as a basis for calculating world production, bearing in mind that the European share of the global volume stands at about 37 per cent. The result was then adjusted by a market/production ratio (a relation between the size of the market and the volume of production). Estimates for 1995 were calculated by referring to the mean annual growth rate forecast for the 1990s. Estimates pertaining to the European multi-purpose vehicles' market were calculated on the basis of recent information taken from the mass media. Hypothesis A, assuming a 3.5 per cent of the share in multi-purpose vehicles production in 1990, has been based on marketing studies undertaken by Matra for the year 1995, indicating a share of 10 per cent in the year 2000. In hypothesis B, numbers for 1990 to 1995 were taken from WirtschaftsWoche, 5 February 1993. The size of the world market in 1990 was calculated by referring to the sum of the European market, the American market (10 per cent of the domestic market at this time) and the Japanese market (identical multi-purpose vehicle share to the European). For the year 2000, the sum of 10 per cent carrier penetration was taken into account, the mean annual growth rate for the 1990s being calculated in order to find out the size of the market in 1995.
Source: CEC, 1992 and the authors' own calculations.

Matra is the European market leader, with about 50 per cent of the market via its production of the Renault Espace. Of the three main USA producers, only Chrysler is manufacturing within Europe. Its Voyager model is produced, to European specifications, in Graz, Austria, in a joint venture with Steyr-Daimler-Puch. The future Ford\VW multi-purpose vehicles will pose a serious challenge for these companies as forecasts predict that they will obtain about a third of the European market, selling between 500,000 and 600,000 vehicles in 1995/6 (Table 9.1).

Matra has already initiated defensive measures to protect its market position, mainly directed at the new Ford\VW vehicle. Its focus is the financial aid and benefits that this joint-venture has been receiving, with EC approval, from the Portuguese government. Direct and indirect aid add up to about 30 per cent of a total investment of ECU 2550 million. According to Matra (Expresso, 31 October 1992) this is unacceptable, for three reasons: (i) it violates competition rules; (ii) it will destabilize the market (the 180,000 vehicles produced at the

AutoEuropa plant in 1997 will swell European production of multi-purpose vehicles to 700,000 while demand has been calculated at 400,000 vehicles); (iii) it is based on a deliberate over-estimate of investment costs in order to be sure of receiving a higher percentage of Community aid. Matra claims that the alleged investment is two and a half times higher than that needed for such a production capacity have, however recently been rejected by the European Court.

The consortium formed by Chrysler and Steyr-Daimler-Puch represents investments worth ECU 300 million, a third of which has been met by the Austrian government. The 1992 annual production of Voyagers was set at 13,000 vehicles, rising to 50,000 from 1994, with 90 per cent of the increase exported to the EC. The level of subsidy has been contested by the EC because it believes inter-company competition will be distorted. The EC therefore suggests that an import duty of between 10 per cent and 20 per cent be imposed because none of the parties involved has been able to reach an agreement.

The European market is currently witnessing tough competition from Japanese producers, through the Mitsubishi Space Wagon, Nissan's Prairie and Serena, and Toyota's Previa. The quality and the reliability of these Japanese multi-purpose vehicles are excellent. The Previa is being manufactured in the USA in order to win a significant share of the world's largest market in multi-purpose vehicles. Forecasts suggest that they could well obtain a larger slice of the European market, however, particularly the Serena because it is manufactured within the EC, at Barcelona.

The fact that the EC is removing restrictions on Japanese products in 1999 will bring new challenges to European auto producers. These will be reflected in the way they prepare their strategies for dealing with increased Japanese competition. In the case of multi-purpose vehicles the Ford\VW project and a planned increase in the output of the Espace are indications of this intensifying competition. Assume that in 1995 the market will be for 600,000 multi-purpose vehicles (Table 9.1), of which European producers would obtain 58 per cent, split between Ford\VW (30 per cent), Renault (20 per cent) and Chrysler (8 per cent). The rest would be taken by Japanese vehicles. Thus if the most optimistic forecasts for market growth are accurate, demand could well exceed "European" production and the shortfall would most probably be met by Japanese multi-purpose vehicles. Nissan and Suzuki have already established production capacity via joint ventures in Spain. If on the other hand market growth is more limited, competition could be much more intense for other manufacturers (PSA and Fiat via a joint venture; Mercedes-Benz) are currently establishing production facilities for multi-purpose vehicles.

The restructuring of the European auto industry will probably continue throughout the 1990s via further centralization of capital and the forging of strategic alliances. The Single Market will favour companies which are part of European production networks, a fact of which most producers had become aware by the end of the 1980s (Sadler, 1992b). This process of "Europeanisation" is particularly observable in the multi-purpose vehicles' segment. In one way, the initiatives of companies producing these vehicles demonstrates the importance of thinking European. The conception and design of the vehicle, involving close collaboration between the main producers and their most important suppliers, suggests that no longer will Europe be simply the sum of different national markets.

The territorial dimensions of production strategies also reveal the importance of "Europeanisation". The integration of the European periphery into the production chains for these important new vehicles is an uneven process. Only the most developed peripheral regions such as Cataluna and Lisboa e Vale do Tejo are being so integrated but this is of great significance for their economies. This at the same time illustrates the emergence of a new map of the automobile production, where the privileged areas of southern European countries are no longer limited to assembling Completely Knocked-Down (CKD) vehicles.

9.3 The Automobile Industry in Portugal

9.3.1 The formation of the industry

Attempts to establish an auto industry in Portugal date from the beginning of the century, based on private sector initiatives. The multiplicity of initiatives was responsible for their limited success. In contrast to other countries with small domestic markets, such as Austria, Belgium, Czechoslovakia, Denmark, Holland, Switzerland and even Spain, which all produced their own particular models, it was impossible to design and manufacture any series of model specific to Portugal (ACP, December 1992). Nevertheless, Portugal's potential market was quite considerable: as well as mainland Portugal, there were also the former overseas colonies.

Failure to found a nationally-owned motorcar industry mainly resulted from the Portuguese government's non-interventionist policy stance. The Estado Novo (New State) completely neglected industrial activity and failed to create the foundations (for instance a national steel industry) on which a national car industry could be built. Later, however, industrial policy changed quite considerably. From the 1960s efforts were made to establish an import-substitution model of economic growth. Access to markets became politically regulated. Severe restrictions upon the import of completely built-up (CBU) vehicles were especially significant. It was decreed that vehicles were to be assembled in Portugal on the basis of at least a 25 per cent Portuguese share of value-added.

This ruling on the characteristics of auto *filières* led to several new assembly lines being established. Because of the variety of models being assembled, however, they were unable to obtain either the necessary scale economies[2] or a stable supply of components from national manufacturers. According to the AFIA (1992), at the end of the 1970s, there were some 21 assembly lines in operation, employing 10,000 workers, and about 170 manufacturers of components and accessories, with 15,000 people working for them. Nevertheless, the previous policy's limitations led to a situation of severe difficulties for most car manufacturers.

In 1979, changes were made in the goals of development policy. The Government now sought to provide conditions to facilitate the auto industry's growth in a market that promised rapid expansion. The sector then came under the jurisdiction of the Motorcar-Industry Act no. 351/79 which, upon enactment, gave the go-ahead for reorganizing the sector. As a result, unprofitable assembly lines were closed down; new, competitive component manufacturers started producing and the volume of exports increased quite significantly.

The most significant result of this new regulatory framework was that, in 1980, it allowed Renault to install a plant, in which the Portuguese state had a minority share. The factory had considerable importance in reviving the economy, even though the crisis had still to run its course. The Renault project led to investment in the components industry of between 45 and 56 million ECU and the creation of 4,000 new jobs between 1980 and 1983 (AFIA, 1992). It is interesting to assess the extent to which Renault achieved its goals as agreed with the Portuguese government in 1980, particularly at a time when the Ford\VW plant is due to open and will be working towards targets that some believe to be over-ambitious. On the whole Renault had a positive impact on the production system. It clearly met some targets, and exceeded others: Renault almost succeeded in reaching its goal for exporting vehicles (45,000 to 46,000 forecast for 1985); it went beyond predictions as regards engines, and produced almost 200,000 gear-boxes and more than 400,000 water pumps that had not been stipulated in the contract. Its goals were practically satisfied as far the foreign-currency balance of payments was concerned. (Público, 9 November 1992).

However, there were other objectives that Renault failed to achieve. It did not attain its direct employment targets, as little more than half of the 6,000 jobs initially forecast actually materialised. Neither did succeed in encouraging local integration, owing to the fragile status of the Portuguese components manufacturers and the progressive internationalization of competition in this sub-sector (see Sadler and Amin, this volume).

---

[2]The Portuguese market consisted only of about 52,000 vehicles (AFIA, 1992).

Table 9.2 Regional distribution of employment in the automobile industry in Portugal

Districts	Total jobs	Jobs on assembly lines	Jobs in component production
Aveiro	5589	603	4986
Beja	123		123
Braga	1890		1890
Braganca	118		118
C.Branco	100		100
Coimbra	368		368
Evora	278	103	175
Faro	211		211
Guarda	540		540
Leiria	1080		1080
Lisbon	7148	1624	5524
Oporto	6403		6403
Portalegre	25		25
Santarem	1845	349	1496
Setubal	3393	1873	1520
V.Castelo	102		102
V.Real	68		68
Viseu	1020	444	576
Total	30301	4996	25305

Source: MESS

9.3.2 The current situation

The auto industry generates 2.6 per cent of the Gross Value Added and 2 per cent of employment in Portuguese industry (GEPIE, 1990). There are slightly more than 30,000 permanent jobs, of which 5,000 jobs are in car assembly plants (ACAP, 1992). Within Portugal, 80 per cent of employment in the auto sector is concentrated in the 5 most industrialized districts - Lisbon, Oporto, Aveiro, Setúbal and Braga. The districts of Guarda, Leiria, Viseu and Santarém appear in intermediate positions, although the first two districts have no assembly plants (Table 9.2 and Figure 9.1).

Two types of vehicle are assembled in Portugal: private cars in the medium/low segment of the passenger-car market (Clio, AX, Astra), sometimes from CKD kits (as with Ford and GM at Azambujca and Citroen at Mangualde: see Figure 9.2); commercial vehicles (Table 9.3). During 1992, 163,000 vehicles were produced, approximately 52 per cent of which were exported. This left insufficient supplies with which to satisfy domestic demand, calculated at about 300,000 vehicles.

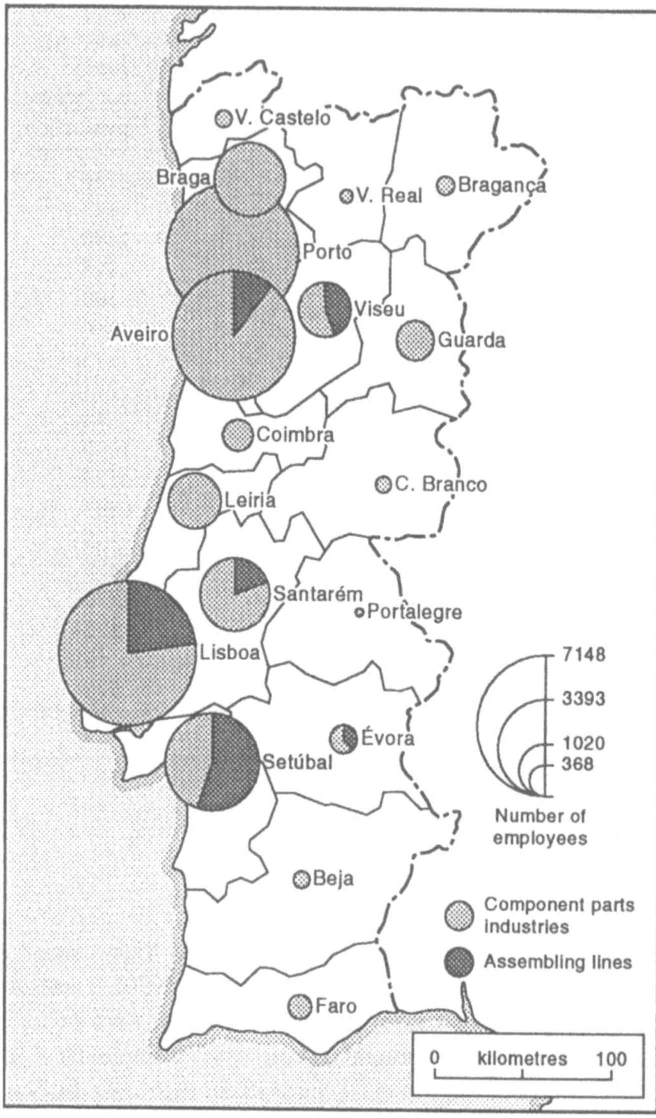

Fig. 9.1 Regional distribution of employment in the automobile industry in Portugal.

The components industry grew quite strongly both in terms of jobs and sales. Employment increased from the second half of the 1980s to around 22,000 in the early 1990s. Sales volumes grew more strongly, with a real annual growth in excess of 10 per cent, while exports grew at a rate of about 20 per cent per year, reinforcing the export orientation of their sector (AFIA, 1992). The structure of the components industry reveals a concentration upon producing

Table 9.3 Assembly lines in Portugal

Assembly Lines	Jobs	Location	Light Passenger Vehicles	Commercial Light	Commercial Heavy
Renault Portuguesa	1403	Setúbal	Clio 19	Clio	
Citroen Lusitana	444	Mangualde	AX	AX	
General Motors	1032	Azambuja	Opel Astra	Bedford Astramax	Bedford
Ford Lusitana	592	Azambuja		Transit Sierra P-Up	
Salvador Caetano	518	Ovar		L. Cruiser Dyna Hiace Hilux	Hino BU96L Dyna Coaster HBD
Movauto	470	Setúbal		Nissan Cab. Nissan P-Up Peugeot 504 Peugeot 505 UMM Alter	Nissan Cabstar
Tramagauto	349	Tramagal		Mitsubishi Pajero Canter FE L-200/300 P-Up	Mitsub. Fuso Canter
Movar	52	Ovar		Mazda 2200 F. 2200 P-Up 3500 ChC.	Mazda T
Batista Russo	103	Vendas Novas	Mini Moke	Mitsubishi L-300	
Soma	33	Ovar			Volvo N/B/NL

Source: Adapted from the ACAP, 1992

engines, transmissions and brakes as well as electrical fittings (27 per cent and 22 per cent, respectively, of total employment: AFIA, 1992).

The new assembly lines, particular those of Renault, have had a catalytic impact on the components industry. Despite its dependency on domestic assembly plants, it is making its presence felt internationally. In future the components sector will increasingly rely on the stimulus provided by the new

Ford\VW plant. According to the company's chairman in Portugal, Renault has helped the Ford\VW joint venture to establish itself there.

9.4 The Ford\Volkswagen (AutoEuropa) Project

9.4.1 Background

Following lengthy talks between the Portuguese Government, Ford and VW, and the Commission of the European Communities (CEC), in July 1991 aid of more than ECU 500 million was approved for the Ford\VW Project, to be set up on the Setúbal peninsula. Endorsement was thus given to a project that would not only leave its mark on the future of the Portuguese economy but perhaps stamp the very geography of the auto industry in southern Europe.

According to its sponsors, the Ford\VW Project is the largest global investment in the auto industry in the last few years - about ECU 2550 million. This will have an enormous impact on the host country: firstly, it is the largest foreign investment ever made in Portugal and sixteen times larger than any other foreign investment made in the preceding two years; secondly, the projected annual exports will exceed those of the textile industry, currently Portugal's main exporter (with about one third of the country's total exports).

Above all else, however, the project must be understood in relation to the restructuring of the USA and European auto industries in response to the increasingly aggressive competition from Japanese producers and the consolidation of their Single European Market (Hudson, this volume; Womack et al., 1990). The companies' strategy in this joint venture is to share risks and the enormous development costs (a third of the total project costs) while winning a substantial slice of the growing market for multi-purpose vehicles. This segment of the market is exhibiting the signs of strongest growth precisely in those regions in which competition among producers in the Europe/Japan/USA triad is at its most fierce. The intercontinental Ford\VW alliance has two particularly innovative aspects. Firstly, it involves two generalist car producers, in contrast to most other strategic alliances and acquisitions and mergers which have involved a generalist company with a specialist producer (such as Ford\Jaguar; Ford\Aston Martin; GM\Saab; GM\Lotus; Fiat\Alfa Romeo; Fiat\Maserati). Secondly, this is the first intercontinental project in which Ford has been involved in which the seat of executive power is not in the USA.

The Ford\VW Portugese partnership can also be seen, however, as a continuation of earlier collaboration between the companies in other environments; for example, they have a long-established collaboration in South America, in Argentina and Brazil (Auto Latina). There are important differences, however. Auto Latina, for example, was formed in 1983 to realize scale economies, widen the range of available models and improve profitability in a market that had previously been difficult for both companies. AutoEuropa, in contrast, is a joint venture specifically formed to focus on an innovative product in fiercely contested markets in the First World.

Would Ford\Volkswagen's decision to manufacture its innovative high-quality product in a peripheral European country such as Portugal not herald substantial changes in the geography of the industry? The Ford\Volkswagen consortium is particularly interesting from this point of view as it involves firms which have revealed distinctive locational strategies (Gaebe, 1991b; Savary, 1991a; Dicken, 1992b). Ford, like other large USA companies in the sector, seeks to reduce production costs; as its products advance along their life cycle, so Ford chooses increasingly peripheral areas. In contrast, Volkswagen selects localities more on the basis of potential sales increases in particular market segments or in regions rather than in relation to the possibility of reducing production costs[3].

However, neither strategy can clearly explain the choice of Portugal for the Ford\Volkswagen project. Ford's "traditional" strategy would suggest the choice of a "central" location for such an innovative product rather than a peripheral one in Portugal. At the same time, however, the small size of the Portugese market and its location on the European periphery are not consistent with VW's "traditional" strategy. In fact, the choice of Portugal reflects its place as a region within the much larger space of the European Union, unfettered by borders and tariff barriers. No longer is competition among national states an appropriate framework within which to understand locational decisions in the auto sector in Europe. The location of the AutoEuropa project in Portugal is a clear sign that automobile production and market have become increasingly Europeanised.

In this case, could we claim - as Savary (1991b) did - that the overall strategies of transnational horizontal integration (product specialization for an international regional market), which tends to replace multi-domestic strategies, generally disadvantage peripheral areas? The innovative character of the AutoEuropa project would suggest that the answer to such a question, in this case at least, would appear to be "no". Nevertheless, such innovativeness contrasts with the traditional nature of both the most decisive factors leading Ford\VW to choose Portugal (and the Setúbal Peninsula more particularly) in preference to other countries, and the position of the Portuguese government, together with the sort of regional political instruments that it has used. In fact, as much in one case as in the other, the strategies which were followed could

---

[3] Despite their differences in strategy, both Ford and VW demonstrate a higher degree of internationalization as compared to their respective American and European counterparts.

not have been more conventional. For Ford\VW, this meant finding ways to reduce costs. This was guaranteed through access to financial incentives and tax exemptions and by low wages in a politically stable environment. At the same time, access to a vast market was assured. As for the Portuguese government, the type of activity it undertook and the kind of approach it selected, strongly reflected the sort of regional policy persued in the 1960s and 1970s. It attracted foreign investments in the hope of modernization, reflecting its pivotal role as a negotiator with big industry, and guaranteed physical infrastructures, incentives and tax exemptions via State policies.

The evident contrast between the innovative character of the product which AutoEuropa will manufacture and the "traditional" character both of the factors that decisively influenced the location of the project and of the means of regional development provide an initial indication of some of the contradictions surrounding this project. From a geographical perspective, it is essential to find answers to two sets of questions:

- does the choice of Portugal, and more specifically Setúbal, represent a divergence from the hierarchical Fordist model of structuring the auto industry's production space in Europe ? Are there ongoing processes which are incorporating some localities in the periphery of Europe into this structure in an equivalent fashion to more "central" localities?
- will this project have local impacts identical to those associated with earlier rounds of branch plant investments? Or will its sheer size and innovative character generate qualitatively distinctive local effects regardless of the reasons that underlay the choice of Setúbal as the location of the project?

Most of the questions raised so far will only find their answers when the project finally gets under way. On-going evaluation will therefore be a means of producing deeper, more precise knowledge about the kinds of sectoral and regional impacts of the project. While the future of the project has been agreed, it has yet to be implemented (work involving clearing the land and erection of buildings was instigated at the end of 1991). The project should begin production in 1995 in a factory of 220,000m^2, continuing until at least 2014. Currently, therefore, it is only possible to summarize the basic characteristics of the project and attempt to make an *ex ante* assessment of its most probable short- and medium-term impacts.

9.4.2 A brief description of the project

Location requirements and conditions. Over 40 possible locations for the plant were considered as part of a very thorough process of site selection. At a certain stage in the project's conception, Portugal and Spain emerged as the most feasible candidates. Within Spain, Sevilla and Murcia offered the most attractive conditions and four possible Portuguese locations were examined (Figure 9.2b):

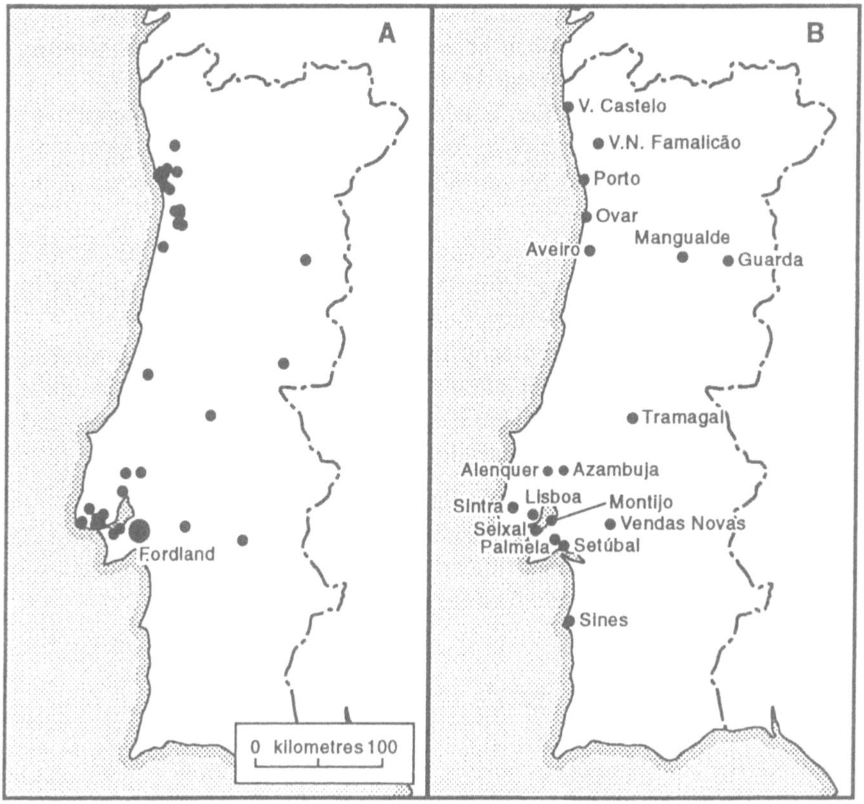

Fig. 9.2 a) Localities of companies already certified as Ford\VW project suppliers.
b) Localities referred to in the study.

- Azambuja, where Ford already has an assembly plant.
- Sines, an industrial port complex dating from the 1970s centred around the chemical industry and connected with oil refining. The Portuguese government wishes to diversify this production structure, which had been in crisis from the mid-1970s.
- Aveiro and Setúbal, port cities that also enjoy good international accessibility by land.

Generally speaking Spain's position as the destination location was prejudiced by its labour market conditions as compared with those offered by Portugal (such as higher salaries and strike rates). Nevertheless, in other respects Spain presented a more favourable environment: for example, the existing development of its auto industry would have been of considerable help when establishing a network of qualified suppliers.

The choice of Portugal and more specifically, the Setúbal Peninsula, mainly reflected the influence of four converging factors:

i) The availability of large financial incentives and tax exemptions, since the Setúbal Peninsula had been declared a crisis area and therefore would benefit from privileged aid awarded by the Community.
ii) An advantageous location between Lisbon and Setúbal, guaranteeing good international accessibility and proximity to both private and public services which would be vital for the satisfactory development of the project.
iii) The availability of an abundant, relatively skilled labour force, with qualifications ranging from the industrial professions to jobs in administration, technical areas and research.
iv) The existence of a privileged locality in terms of the environment, combining the advantages offered by proximity to a large urban centre (for example, access to higher quality social and cultural resources) and of living and working in scenically attractive rural surroundings.

With the exception of the first factor - which in this case was decisive - the criteria used in choosing the site tended to follow the general trend witnessed among new industrial plants funded by foreign capital and possessing an administrative and/or an R&D section (Savary, 1990).

9.4.3 Investment and Financing

In all, the Ford\VW project eventually attracted ECU 750 million of aid, not all from the ERDF, distributed as follows:

i) **Financial Incentives.** These amounted to about ECU 500 million, of which 70 per cent was guaranteed by the ERDF as non-reimbursable, and the remaining 30 per cent by the Portugese government. The European Regional Development Fund (ERDF) funds form one part of the allotment of structural funds to Portugal from the EC. The

Table 9.4 Impact of the SIBR incentives system (1988-1991)

	Projects		Investments		Incentives	
	Total	Per cent	PTE 10^6	Per cent	PTE 10^6	Per cent
Setúbal district	1(Ford\VW)	0.1	453.0	30.5	89.1	29.0
	155	9.1	140.2	9.5	27.8	9.0
Subtotal	156	9.2	593.2	40.0	116.9	38.0
Other areas in mainland Portugal	1540	90.8	891.4	60.0	190.8	62.0
Total	1696	100.0	1484.6	100.0	307.7	100.0

Source: Direcção Geral de Desenvolvimento Regional - authors' calculations.

importance of the Ford\Volkswagen project in relation to the total amount of incentives between 1988 and 1991 that were channelled through the Regional Incentives System (Sistema de Incentivos de Base Regional - SIBR - which in turn obtained its funds from the ERDF) is easily shown (Table 9.4).

ii) **Tax Exemptions**. These totalled some ECU 50 million over a five-year period.

iii) **Aid for Vocational Training**. This was in the order of ECU 203 million drawn from the European Social Fund programmes and already included in the Community's Aid Framework forecasts.

At the same time, the Portuguese government undertook to provide the area with substantial infrastructure investments, of about ECU 102 million. This would involve building new roads and railway-lines (including a siding going to AutoEuropa and linking it to the port of Setúbal), a waste-treatment station, a roll-on/roll-off portside terminal and an occupational training school for the auto industry.

9.4.4 Production and Markets

By 1995, annual production of the 7-8 seater multi-purpose vehicles (currently designated as the VX 62) should stabilise at 180,000 vehicles - 820 vehicles per day in three shifts. This will be about a third of the total European market for such vehicles. At present (1993) 160,000 vehicles are assembled annually in Portugal, a comparison which emphasises the significance of the project for that country.

Of the 180,000 vehicles, about 162,000 will be exported by sea and the remaining 18,000 overland. The Portuguese market would only absorb 1 per cent of the production which is destined mainly for the European market (Germany, the UK and Spain in particular). The vehicles will be differentiated by their own distinctive marques and variations in style and distributed in identical proportions by the Ford and VW sales networks to comply with the Community's anti-trust legislation. The existence of completely separate marketing networks, managed on data bases that are unavailable to the other partner, is indicative of the complexities of the competitive/cooperative strategies that major companies must now develop.

The relative proportion of national added value was set at 40 per cent up to 1998. After this, the Portuguese component manufacturers involved in supplying the project would contribute 45 per cent of the added value; there would be an additional 10 per cent if sales were made by domestic producers to other Ford\VW plants in Europe. It has been calculated that in the next few years, this project will induce a 70 to 80 per cent growth in the components sector in Portugal (assessed at ECU 1,580 million in 1991). This will involve a total investment of ECU 847 million by both Portugese and foreign suppliers

Table 9.5 Profile of direct employment to be created by the Ford\VW project

Professional qualifications	Employment
Management and technicians	240
Department heads/foremen	180
Skilled workers	600
Non-skilled workers	4000
Total	5020

Source: Ministry of Industry and Energy

located in Portugal. A Government department specifically to provide aid to Portuguese companies involved in this project was set up on July 24 1991 under the auspices of the Portuguese Institute for Aid to Small and Medium-sized Enterprises (IAPMEI). Its top priority is to find and choose companies that would supply components. The IAPMEI plans to allocate ECU 26 million to small and medium-sized suppliers in their specific endeavours to back up the Ford\VW project.

9.4.5 Employment

There is little information available on this subject. Calculations indicate that 5,000 direct jobs (see Table 9.5) and about 7,000 indirect ones will be created in the region. Wages and salaries will, however, only account for about 0.8 per cent of the total production cost. Recruiting personnel started a little while ago and has included the transfer of Portuguese emigrants living in Germany or the USA[4] who are employed as technicians or skilled workers in other Volkswagen plants (in Hannover for example) and Ford factories. These Portuguese workers will train the first teams of workers. Furthermore, it is planned to bring in higher management and engineers from other Volkswagen and Ford plants in Europe.

From a qualitative point of view, the sums which are necessary to train the available labour force (and about ECUs 215 million have already been set aside by the Portugese government for professional training programmes) give an idea of just how much effort is needed to re-train personnel, even though most of them will be unskilled (Table 9.5). Formauto, the occupational training school, was due to go into action. Built inside the AutoEuropa grounds, it has been organized in partnership with the Institute of Employment and

[4]According to the press (Expresso, 20 February 1993), AutoEuropa "sent letters to all the U.S. consulates asking them to sign on immigrants who wished to return to Portugal".

Table 9.6 Firms in Portugal classified Q1 by Ford

Name	Activity	Location
Acumuladores Autosil	Batteries	Oeiras
Bendix Portuguesa	Brakes, Components	Abrantes
Bertrand-Faure Portugal	Seats	S.J.Madeira
Cablesa (G.M.)	Electric Cables	Cascais C. Branco Oeiras
Continental Mabor	Tyres	Famalicão
Covina (St. Gobain)	Window Glass	S.I.Azoia
Gametal (Bertrand-Faure)	Stamped Metal Parts	O.Azemeis
I.A.C.M.	Stamped Metal Parts	Matosinhos
Impormol	Springs	Azambuja
Indelma (Siemens)	Electric Cables	Seixal
Manuel Conceição Graça	Stamped Metal Parts	Alenquer
National Starch Chemical	Chemical Products	Oeiras
Pereira Barroso Oliveira	Stamped Metal Parts	V.N.Gaia
S.I.M.G.	Stamped Metal Parts	Guarda
Siemens	Relays	Évora
Sonafi	Steel and Cast Iron Parts	S.M.Infesta
Tavol	Stamped Metal Parts	S.J.Madeira
Teledinâmica	Control and - Transmission Cables	Maia
United Tech. Automotive	Electric Cables	Valongo
Yasaki Saltano Portugal	Electric Cables	V.N.Gaia

Source: IAPMEI (1993)

Table 9.7 Firms in Portugal classified Q101 by Ford

Name	Activity	Location
Arjal	Stamped Metal Parts	V. Novas
Fapobol	Rubber Parts	Porto
Funfrap	Steel and Cast Iron Parts	Aveiro
Inapal	Stamped Metal Parts	S.M.Infesta
Map	Plastic Parts	Leiria
Simoldes Plásticos	Plastic Parts Thermo-Moulded Parts	O.Azemeis
Sotinco	Paints and Varnishes	Barreiro
Trecar	Inner Coverings/ Linings	S.J.Madeira

Source: IAPMEI (1993)

Professional Training. The centre will provide training to any company working within the auto industry. If more specialized training is required, the programmes at the centre will be backed up by apprenticeships in other Volkswagen and Ford plants within Europe.

Data on key issues, such as profiles of required qualifications, is unavailable. Information has only been given regarding the type of organization of labour, which will imitate the "Japanese model" of team working.

9.4.6 The process of setting up supplier-networks

Establishing a network of suppliers is mainly Ford's responsibility. Portuguese companies which become direct suppliers will be chosen on the basis of Ford-awarded certificates of quality. At the end of January 1993 (according to IAPMEI-Gapin), 20 companies had been classified as Q-1 (guaranteed suppliers) and 8 as Q-101 (potential suppliers) (Tables 9.6 and 9.7). Generally speaking, these companies, which included some plants backed by foreign capital and producing in Portugal, manufacture relatively simple (Type C) components. By mid-1993, it is expected that 60 firms will receive a Q-1 quality certificate.

Setting up a joint-venture is warranted in cases where a Portuguese company has difficulty in securing its status as a supplier of more complicated components (Type B - medium complexity - or Type A - high complexity)

Table 9.8 Ford\VW suppliers to be located in Fordland

Name	Activity
Alfred Teves	Brakes
Benteller	Springs, Shafts
Edscha/Sharwachter	Pedals
Gillet	Exhaust Pipes
Hohe	Rear-View Mirrors
Inapal-Menzolit	Composite Parts
Kickert	Locks
Palmetal	Stamped Metal Parts
Rockwell Golde	Roofs
Sikamoldes (Simoldes/Kautex)	Plastic Parts
Sipavel-Unikeller	Cabin Silencers
T.M.G./Allibert[1]	Bumpers Control Instruments Inner Coverings/ Linings
Vanpro (Johnson Controls/ Bertrand Faure)	Seats

[1] Joint-Ventures not yet formalized

Sources: Expresso, Expansão, Público (various dates).

Table 9.9 Other Ford\VW suppliers

Name	Activity
Potential Suppliers	
Allibert	Mats
Autoliv (Electrolux)	Seat Belts
Caourepe	Rubber Parts
Dalphi Metal Portugal	Driving Wheels
David Valente Almeida	Stamped Metal Parts
Ford Electrónica Portugal	Control Instruments Radios
Huf Portuguesa	Locks
Laboratorios Radio	Composite Parts
Lemmerz	Wheels
Nadella	Direction Bars
Robbialac	Paints and Varnishes
Rockwell International	Springs, Shafts
Ronal	Wheels
Salvador Caetano	Stamped Metal Parts
Joint-Ventures	
Fapobol-Diehl	Rubber Parts
Ibermoldes-Olhotechnik	Plastic Parts
Impormol-Hoecsh	Springs
Ipetex-Viktor Achter[1]	Inner Coverings / Linings
Kuntz\Karnein-Facort	Machining
Moltec-Sherdel	Springs
Plastifil-Dynamit IB. Nobel	Plastic Parts
RTJ Plásticos-Riesselmann	Plastic Parts
T.M.G.\De Witte	Inner Coverings/ Linings

[1] Joint-Ventures not yet formalized
Sources: Expresso, Expansão, Público (various dates).

without resorting to sub-contracting abroad for parts or services that are not produced in Portugal. Different joint-ventures, some with a majority share of Portuguese capital and involving partnerships between Portuguese firms and Ford and Volkswagen's international suppliers, have already been formed or are being formed. The commodities that these joint-ventures produce reflect the specificities of the Portuguese industrial structure: injectable and inflatable plastics, cables, fabrics (for upholstery coverings and inside car-linings), window-glass, etc. (Tables 9.8 and 9.9). In some areas, such as the foundries, joint-ventures have been difficult to set up due to the poor quality of potential Portuguese partners.

There are two more factors of relevance to Portuguese companies becoming direct suppliers to AutoEuropa. On the one hand, some "Portuguese" firms which are large enough to apply individually are, in reality, controlled by foreign capital. In fact, the rigorous demands (in relation to techniques, product quality control, efficient management, etc.) imposed by Ford\Volkswagen form unsurpassable barriers to the overwhelming majority of exclusively Portuguese-owned companies. On the other hand, even when there are no restrictions as regards the quality of the products, Portuguese companies' size and organizational structures could create serious difficulties. For these firms, being Ford\VW's suppliers means depending almost exclusively on a single market; medium and long term, this could turn out to have extremely negative consequences[5].

It is likely, therefore, that the 40 per cent share which Portuguese industry is expected to contribute to the project's added value is not only half fictitious (it is not accidental that joint-ventures should preferably have a majority Portuguese involvement), but, as with Renault, it is also likely that this share will not be fulfilled within the next few years.With regard to foreign companies, it is clear that the relations which Ford has already set up internationally with its suppliers, will tend to be reproduced in Portugal and will play an important role in structuring the emerging linkage network. In fact, some foreign companies, which supply Ford, are already investing in Portugal or have announced their intention of doing so soon. These firms will fill gaps in the Portuguese production system, investing in segments such as ABS brake systems and electronic parts, and light alloy wheels.

Some foreign companies which are the Ford\VW project's direct suppliers are seeking to extend their control of their respective value added chains in Portugal. For example, Continental (Germany) already controls the Portuguese market in tyre production and it is now attempting to control the market for tyre supplies by buying a Portuguese company which is well- placed in the activity. This process, involving vertical integration in relatively specific spheres of activity in a small country like Portugal, allows for a succession of markets to be established without the risk of domestic competition.

Finally, as this is only the first phase of the formation of a supplier network to AutoEuropa, nothing can be said about the indirect suppliers, those companies producing under a system of sub-contracting for other firms that have been selected as direct suppliers. Nevertheless, the number of indirect suppliers is likely to be high because Ford\VW are looking for companies which can supply component systems rather than isolated components.

---

[5]It has been calculated, for example, that 25 per cent of the tyre production in Portugal (about 4 million) will be channelled to AutoEuropa (Diário de Notícias, 21 September 1992).

9.4.7 The geography of direct suppliers: possible outlines of an emerging archipelago

The different situations in which the direct Ford\VW project suppliers find themselves, and their adoption of just-in-time (JIT) principles of production and management, have aptly demonstrated the way in which a network, which is extremely demanding in meeting its deadlines, has come to rely on physical proximity for its efficient functioning.

Different authors (for example, Paché, 1991; Veltz, 1992) have, however, pointed out that the JIT principle is not necessarily associated with dense spatial concentrations. The ability to deal with time constraints could be much more important than guaranteeing closeness at hand: "deadline proximity" overshadows "physical proximity" so that the transaction space of a given production network does not need to be restricted to a certain geographical area. Giunepero and O'Neal (1988) point out that the "Instrument Division of the U.S.-based Davidson Rubber Company gets supplies to the industrial plants working in the motorcar trade in Detroit four times a day despite the fact that it is located 1,500 km away" (cited in Paché, 1991, 66). Lung (1992) has also noted that adopting Toyota's approach of the flexible organization of production does not necessarily require spatially concentration.

However, taking an opposing view grounded in a more or less sophisticated extension of the Marshallian concept of "industrial districts", others have stressed the advantages of forming territorial clusters. In these, physical proximity becomes a crucial factor both for reducing all kinds of transaction costs and, above all, for intensifying and qualifying production processes and the circulation of new, specialized information (see, for example, the summary in Storper and Harrison, 1992).

The Ford\VW project does not appear to correspond exactly to either of these positions. On the one hand, having direct suppliers near at hand was thought to be essential: next to the AutoEuropa plant is Fordland, 21 hectares of land that could be extended in the near future. The direct suppliers are encouraged to locate here, with either production plants or warehouses holding enough stocks to meet previously- agreed deadlines[6]. Nevertheless, the Ford\VW cluster will not probably function as a modern industrial district, a localised production system that stimulates externalities and synergies influencing the local entrepreneurial fabric. There are various reasons for this. First, Fordland mainly houses foreign companies, which have established themselves for the first time in Portugal, and warehouses belonging to factories

[6]This strategy could have something to do with reducing the risk of factory workers' striking. The recent strikes in Renault's engine plants in Cleon had serious consequences for the company because they affected its JIT system.

situated in other parts of the country (Table 9.8). Secondly, the number of firms which have already set up in the area and been selected as direct suppliers is fairly low. Thirdly, the Setúbal Peninsula is so close to saturation point that public institutions have been advising investors to locate their projects elsewhere in the country. Fourthly, some foreign investors have been particularly responsive to increasingly attractive conditions being offered them by different municipalities in order to attract further outside investments[7].

The situation arising with the Ford\VW project would surely not correspond to the extreme case of a territorially disintegrated quasi-vertical integration. However, it would also be inappropriate to think of it as a case of territorially integrated quasi-horizontal integration. In terms of the typologies put forward by Storper and Harrison, 1992 (278 and 287) which define the basis on which predominant governance and input\output systems operate in each of these cases, we could perhaps say that the Ford\VW project is situated halfway between Boeing\Seattle and Toyota City. The latter both present core-ring hierarchical systems in which the lead company strictly circumscribes the activities of its suppliers, on the basis of asymmetrical power relationships. While hierarchical, however, these relationships are forged over widely differing geographical areas, ranging from dense localities (Toyota City) to dispersed areas (Boeing\Seattle).

In the case of AutoEuropa, a Ford\VW archipelago could very well evolve, with the following features. First, a main core which basically coincides with Fordland, centred around AutoEuropa, and forming a cohesive, extroverted enclave. Secondly, a ring with a 60-80 km radius measuring outwards from AutoEuropa, benefitting not only the immediate proximities of the plant and the various direct suppliers located in Fordland, but also favouring places near to Lisbon because of externalities and international accessibility. Thirdly, a few small nuclei which are both the result of successful deals made between some foreign companies and local authorities[8], and the formation of clusters around

---

[7]For example, OPTEC DD - an English company associated with Mitsubishi - will be investing about ECU 12 million in a very advanced technological plant producing ignition systems, cables and electronic components for the motorcar industry. The company is already supplying ignition systems to Ford in the USA and it is now seeking to do the same thing as regards the Setúbal project. Despite this, it has settled on a site in the industrial zone of Viana do Castelo which is situated near the northern border with Spain (Galicia) and more than 400 km away from Setúbal.

[8]It was recently announced that various Spanish enterprises (or subsidiaries of multinationals, such as Electrolux) - all of which are Ford suppliers - would be setting up business in Vila Nova de Cerveira (more than 400 km away from Setúbal). Although it was planned that these companies would operate on the Setúbal Peninsula, they all responded positively to the incentives offered them by the local authorities and preferred the give-away price of land in the north to the advantages of being located near the Ford\VW plant.

some of the direct suppliers which frequently seek the services of sub-contracting firms. Fourthly, sites of already existing companies which were chosen to be AutoEuropa's suppliers, working either alone or in joint-ventures with foreign companies.

Although still few in number, the companies already selected as guaranteed direct suppliers (Q-1 quality certificates) and potential suppliers (Q-101) could suggest the eventual outline that the Ford\VW archipelago will take (Figure 9.2a). In the future, it will be of the utmost interest to perceive how this intra-sectoral archipelago will be articulated with traditional territorial clusters that have been previously developed on sectoral specialization patterns.

9.5 Concluding Comments

Despite the fact that it is still in its initial phase, the geography of multi-purpose vehicles' production in Europe would appear to favour some of the most developed regions in the peripheral countries: Ford\VW in Setúbal (Portugal), Nissan in Barcelona (Spain) and Chrysler in Graz (Austria). This trend seems to reveal a selective process of integration of specific fringes of the European periphery into global networks, one part of a process of "Europeanisation".

The innovating, expanding nature of this segment of the auto market leads to the likelihood of projects such as these being regarded as able to upgrade the quality of the productive fabric of the host areas. The reason for this is due to the formation of a wide network of suppliers, selected on the basis of a demanding certification process. On the other hand, multi-purpose vehicles' production could help to consolidate intra-sectoral specialization patterns that are opposed to the traditional inter-sectoral patterns prevalent in most of the peripheral regions.

The production of multi-purpose vehicles would therefore appear to open up new possibilities for these regions, helping them to make a qualified integration within the specialized intra-sectoral scenario which characterizes the most developed European regions. Nevertheless, analysing the Ford\VW project reveals the difficulties, tensions and contradictions that emerge from such an initiative in a small semi- peripheral country. The fact that the project is still in its launching stage makes it an excellent observatory precisely because it has still to take off. Analysts will, therefore, be able to follow the way in which global strategies, responsible for restructuring the auto industry within a highly competitive environment from which innovating segments are emerging, are capable of providing new opportunities for some of the peripheral regions or, on the contrary, are creating the foundations for new and more sophisticated cathedrals in the desert within a fractured Europe.

Chapter 10
Interdependent and Uneven Development in the Spatial Reorganisation of the Automobile Production Systems in Europe

Ray Hudson[1] and Eike W. Schamp[2]
[1] Department of Geography, University of Durham, United Kingdom
[2] Institute of Economic and Social Geography, Johann Wolfgang Goethe University, Frankfurt am Main, Germany

10.1 Introduction

The preceding chapters have drawn an iridescent picture of different automobile production systems in varying states of change and dissolution in a variety of locations in Europe. There is ample evidence both of continuing fragmentation of markets and production and of integration on a broader European scale taking place simultaneously. In the introductory chapter of this book, the broader production system approach was suggested as a promising framework for gaining an understanding of the ongoing changes in the spatial organisation of automobile production of Europe. Although this approach has been used more implicitly than explicitly in several of the subsequent chapters, the analyses in these chapters have placed particular emphasis upon the basic changes currently at work in the transformation and regulation spheres of production (see Fig. 1.3). In this chapter, we will attempt to tie together, albeit imperfectly, the different strands of the story in order to achieve a more comprehensive understanding of contemporary changes in the spatial organisation of the emergent European automobile production system.

First, however, some particular factors, which have often been previously neglected but which influence the anatomy of automobile production systems, will be briefly considered. In drawing the potentially new map of automobile manufacturing in Europe, it is important not to neglect the influence of spatial

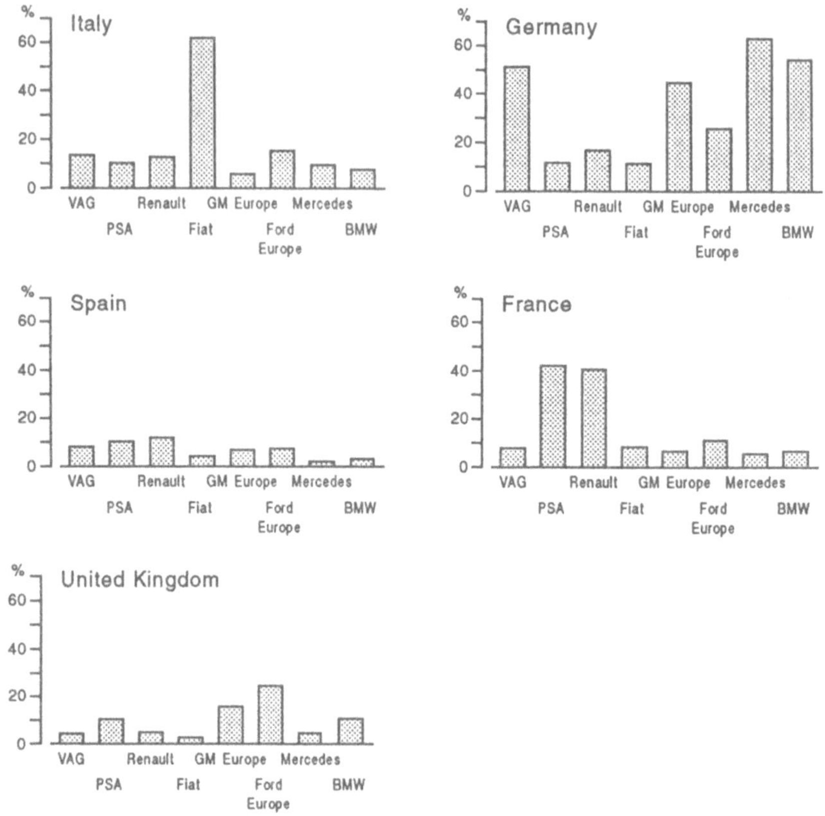

Fig. 10.1 Automobile 'champions' on the major national markets in Europe, 1993
Source : Bordenave and Lung, 1993, page 45

patterns of distribution and consumption. For markets remained - and indeed still remain - regionally differentiated and diversified. This is partly because of variations in consumer preferences and household income levels, partly because of different distribution strategies of automobile producers, and, not least, because of continuing differences in state regulatory frameworks affecting car consumption, such as tax laws. In these ways, "local" resistance to tendencies towards globalisation of production and products effectively continued. Thus, the largest European automobile producers still remained 'national champions' in their domestic markets, as is shown in Fig. 10.1. Smaller national markets such as the Netherlands, Austria and Switzerland have been shared between some of them for a long time, particularly between Volkswagen and Opel. Typically, the larger markets form the core area of traditional automobile production and distribution in Europe. There are, however, two other differing cases in Europe. One is the United Kingdom which experienced a complete breakdown of its national champion during the

1970s and a revival of its automobile production system via Japanese transplants during the 1980s (see Hudson, this volume). The other is Spain which experienced a rapid acceleration of motorization and automobile production via 'European' transnationals - including Ford and Opel in particular - during the 1980s. In both cases expanded production within their respective national territories is mainly bound for export to the European core regions.

National markets have so far not been exposed to the turbulence and further differentiation that undoubtedly will follow deeper integration within the European Union. Japanese companies, for example, still remain excluded by continuing restrictions on production and trade as even internal barriers are only hesitatingly dismantled (see Sadler, this volume). Nevertheless, demand is becoming diversified horizontally via the introduction of new automobile products in largely saturated markets. Although there is an obvious production over-capacity in the automobile sector in Europe, new production sites have emerged - and continue to emerge - but before others are fully closed.

Explaining changes in the production systems merely by strategic decisions of actors such as firms, trade unions and governments can result in the importance of macroeconomic conditions being underestimated. The potential cyclical effects of economic upturns and downturns is one case in point. Of perhaps greatest significance, however, are alterations in currency exchange rates. These can be of tremendous relevance in shaping the long-term strategies of enterprises. They influence both comparable costs of production and prices of products on export markets. As most European automobile producers send the bulk of their exports to other European countries (see Chapter 1), the European Monetary System insulated firms from such impacts during the 1980s. But the emergence of a group of hard currency countries which belong to the production core in automobile manufacturing, namely France and Germany together with their more or less dependent currencies such as the Belgian Franc, the Dutch Guilder and the Austrian Shilling, led to a dichotomy within the European economy. The ability of Volkswagen to export the Golf model to Italy and the competitiveness of German component suppliers suffered heavily from the exit from the European Monetary System and the strong devaluation of both the Italian and the United Kingdom currencies in 1992. In contrast, automobiles from Japanese transplants in the United Kingdom gained price advantages in continental Europe. The devaluation of the Swedish Crown in 1993, which was one reason for the restored profitability of Volvo in its exports markets, may have contributed significantly to Volvo's withdrawal from its proposed merger with Renault (see Savary and Malmberg, this volume). Although in the long-run there is no necessary deterministic relationship between currency exchange rates and international competitiveness (for instance, advantages stemming from devaluation and undervaluation may be reversed by domestic inflation rates), huge differences in currency exchange rates are of prime importance at present in shaping changes in automobile production systems in Europe. Consider two examples; first, the undervalued currencies in central and eastern Europe which make supplier production competitive compared to southern Europe; secondly, the global revaluation of the Japanese yen which stimulates Japanese automobile producers to direct foreign investments. As a result, processes of further decentralisation of production from the hard currency and traditional core areas to the soft currency and peripheral areas in Europe will certainly continue (Bordenave and Lung, 1993).

If markets largely remain 'local', the question then arises as to how market access can be ensured in the context of these decentralizing tendencies in the geography of automobile production. Maintaining this is the domain of new forms of circulation which enable firms to maintain reliable, quick and cheap access to 'local' markets from very distant production sites. Several examples have been given in this volume which point to new modes of transportation and developments in logistics: multi-purpose vehicles made in Setúbal in Portugal will be mainly transported to their west European markets by ship (Ferrão and Vale, this volume), metal stamped parts are delivered from Opel's plant in Zaragoza to the new Eisenach plant by train (Schamp, this volume), and the first clients of the new Eurotunnel between the United Kingdom and the continent, which was opened in July 1994, are reported to be automobile producers such as Ford, Rover and PSA and component suppliers transporting parts. This could encourage components production within the United Kingdom, following the example of Bosch's plant at Miskin, Cardiff (Sadler and Amin, this volume).

All this could be interpreted within a neoclassical framework, that is as reflecting exogenous changes in market conditions to which firms have to adjust in various ways, including via locational choices. Even locational choice, however, is not simply a reflection of the workings of unfettered market processes in Europe. Not least, these are never simply and only locational choices but unavoidably involve linked choices about product, process and markets. There is, furthermore, considerable evidence of unequal bargaining power between states and large firms and between 'small' supplier firms and their large clients in the determination of locational choices. The highly influential MIT study (Womack et al., 1990) has not only been criticised for its suspected errors in the calculation of productivities (Banville and Chanaron, 1991; Savary, this volume) and its static view but also particularly for neglecting the political-economic context in which automobile production takes place. There are numerous reports about bargaining between automobile producers and their component suppliers over 'follow-the-client' locational decisions. Another dimension of this context is the impact of state subsidies on locational decision-making and the emerging conflict between the European Union Commission and national governments within Europe over public subsidies, as exemplified in various chapters of this book (Hudson, Ferrão and Vale).

The Commission's regional policy increasingly seems to strengthen tendencies towards the decentralisation of production within Europe. Combined with the introduction by companies of new concepts of 'lean' production, this could result in a considerable reduction of investment costs in greenfield sites in the fringes of Europe and, conversely, a tendency for a relative increase in the wage cost disadvantages of traditional core areas. Such subsidies may result in investments in peripheral regions which, seen from the region's perspective seem enormous, as is shown with the Setúbal example (Ferrão and Vale, this volume) or with Skoda in the Czech Republic. Seen from the firm's perspective, however, the reduction of investment costs in greenfield sites reduces the importance of 'sunk costs' as a barrier to future disinvestment and, hence, enhances plant mobility. This consideration may have influenced decisions by automobile producers to allocate risky production of low and medium volume new products such as multi-purpose vehicles and off-road vehicles to the European fringe. For they have been located there even though such new models need proximity to markets which, particularly for these new products, are in wealthy regions; that is, in the European core.

Consequently changes in the dominant concept of automobile production in Europe from Fordism to something else (variously denoted as Toyotism, 'lean' production, flexible automation, mass customisation), a change necessitated by the global competition of Japanese automobile producers, are embedded in a context in which systems of circulation and modes of regulation remain regionally and nationally distinctive. It seems as if the debate about transferability of Japanese production concepts to Europe has finally come to the conclusion that Japanese principles work in Europe as well (Dankbaar 1992). Some French scholars, however, have claimed that a particular adjustment of those production principles to the European socio-political environment is necessary, giving rise to what they have called a 'hybridisation' process of production concepts (Lung and Mair 1993), something that has also been recognised by managers in these new Japanese transplants (see Hudson, this volume). Sadler (this volume) has pointed to the ambiguous current situation, hovering between preserving national environments and establishing a new common socio-political environment for automobile production in Europe. Nationally - if not regionally - bound regulatory frameworks for automobile production clearly still persist in Europe. Such a conclusion suggests a need to consider, even if only briefly, the macro-geography of automobile manufacturing in Europe and possible changes in it.

There is an increasing debate about 'the' emerging European automobile production system and, hence, given the more or less slow dissolution of national production systems, something new (Jones, 1993; Bordenave and Lung, 1993). Traditional geographical concepts, such as centre-periphery or spatial divisions of labour, seem, at first glance, to be useful for understanding the spatial order of changing automobile production systems in Europe. In fact, automobile and component production is still geographically concentrated in a North-West-European triangle, considered earlier by Jones and North (1991) and recently re-configured (Jones 1993). At the same time, this region contains the largest national markets, characterized by changing demands posing challenges to automobile producers. Demand and production seem, however, to be growing more rapidly in what has been called the European semi-periphery and periphery; that is, in the United Kingdom, Spain and, very recently, in eastern Europe. Although this seems to be an over-simplified model of the emerging spatial organisation of automobile production in Europe, it provides a useful initial starting point from which to summarise current tendencies in the sector. Broadly speaking, three major tendencies will influence the future spatial organisation of automobile production in Europe:

- new production sites in new locations, mostly in the periphery or semi-periphery of Europe, will threaten the traditional core regions;
- the future role of central and eastern Europe as both market and production region still remains unclear but may have influences upon southern and western European fringe locations;
- in traditional core regions new forms of resistance against the relocation of production to the European fringes are emerging. This may result in an 'upgrading' of functions in core regions and a further deepening of the hierarchical relationship between the core and those fringes.

What kind of spatial re-organisation will these processes produce? Will there be a new map of automobile production in Europe? If so, what will be its characteristics? It is to a consideration of these kinds of questions that we now turn.

10.2 Challenges to the Core from Western and Southern Europe

The 1980s have seen a major shift in automobile production in Europe, both to the United Kingdom and Spain. Assuming that, then, the core of automobile production in Europe still consisted of different national production systems, the (re-)emergence of production in these two countries contributed to the undermining of those systems. It represented the first tentative steps towards a new European production system. The distinctive new features of this emergent European system can be summarized in three key concepts: first, *Japanisation*, either in the form of Japanese transplants being established in the United Kingdom (see Hudson, this volume) or in the form of new production concepts introduced into new greenfield sites; second, *Europeanisation* either via the emergence of new production sites for new products such as off-road vehicles or via relocation of small-sized car production; and, third, a first wave of *relocation of automobile suppliers.*

Japanese producers such as Nissan and Suzuki already had started production (of off-road vehicles) in Spain in the early years of the 1980s. It was not, however, until Nissan established its transplant at Sunderland that Japanese production in Europe was considered to be a major competitive threat to the European producers. Perhaps the diverse experiences with transplants in the United States of America were the reason for this reaction (see Kenney and Florida, 1993). In fact, transplants production was oriented, first, to volume production of medium-sized automobiles, which formed the core market segment of most European companies. The Nissan factory had several important competitive advantages. It was based on the deployment of new production concepts on a greenfield site. It recruited a young workforce, which allowed a considerable intensification of the labour process and high levels of labour productivity. Moreover, it received generous financial support from the state in a generally supportive regulatory environment. Consequently, Nissan's Sunderland factory was expected to be the forerunner of further Japanese transplants which would seriously erode the traditional markets of continental automobile producers. Obviously, however, Japanese producers increasingly focused on proximity to more traditional automobile producing locations for such medium-sized cars, such as Burnaston and Swindon for Toyota and Honda, respectively. Latecomers to Europe, such as Mitsubishi and Mazda, have preferred take-overs or cooperation with peripheral plants in the European production core as the route to establish a production presence within Europe: Mitsubishi plans to produce a new model in the Nedcar plant at Born in the Netherlands from 1995 on and Mazda, which already has cooperative arrangements with Ford in other parts of the world, negotiated to produce a model in Ford's Saarlouis factory. Those plants, however, differ in one crucial respect from the main plants of European automobile producers. They lack R&D capacities and, hence, depend largely upon R&D centres abroad.

Whereas the transplants increased competition in the segment of medium-sized models, the new locations of automobile production in Spain and southern Italy have squeezed the segment of small-sized models and occupied the segment of special models. Lagendijk (1994) has analysed the evolution of the emerging Spanish automobile production system from early import substitution to export orientation and, during the 1980s, full integration into the European automobile system. When the Spanish government switched to liberalisation policies in the 1970s, Ford Europe was the first to start a truly European production network via establishing a greenfield plant for production of the Fiesta at Valencia in 1976. In the same year, Renault took over the Valladolid plants (now under the name of Fasa Renault) in which it currently produces the small Clio and Twingo models (see Savary, this volume). Opel followed with its greenfield plant at Zaragoza for production of the small-sized

Corsa model in 1982. Finally, VW acquired Seat, mainly to produce the small Ibiza and Polo models (see Schamp, this volume). Production of small-sized models has become mainly export-oriented, although the Spanish domestic market grew quite strongly during the 1980s. There were two main reasons for relocating small-model production to the Iberian peninsula: penetration of rapidly growing domestic markets and lower costs of a qualified workforce, compared to the European core. Comparative cost advantages are not only determined by productivity - which has been fairly good in Spain - but also by currency exchange rates and domestic inflation. It is in this realm that automobile production in Spain faced some problems recently.

In addition to receiving these investments, the fringes around the European production core have been chosen for production in the emergent small volume segment of special models, such as cross-country vehicles and multi-purpose vehicles. In the early 1980s, Nissan took over the plant in Zona Franca in Barcelona for the assembly of different models, including recently the Terrano model of Opel; Suzuki acquired Santana at Linares, in Jaen, for assembly of the Vitara model. There are, however, other competitive locations for cross-country vehicles assembly in Europe: in 1994, Chrysler started to assemble the Grand Cherokee model at Graz, in Austria (for export to the European Union). This is a joint venture with Steyr-Daimler-Puch, begun in 1992 with a greenfield site for the assembly of Voyager multi-purpose vehicles. Those assembly plants are often characterized by an absence of metal stamping facilities and, more generally, by a low degree of automation, leading to a reduction of investment costs and limited production capacities (40,000 to 50,000 vehicles per annum). Once more currency exchange rates have become important in the calculation of profitability as 'lean' production sites depend upon large inflows of component supplies. If these are mainly imported, as was the case with Seat and will be the case with AutoEuropa in Portugal, the consequences of exchange rate changes for cost management are quite obvious and can be very serious.

The new growth segment of the multi-purpose vehicles or people carriers was, for a few years from 1983, occupied only by Matra with the Renault Espace model. It is now increasingly coming under pressure from new competitors, with assembly sites in both northern and southern Europe. These chose a risk minimizing strategy in the organisation of joint ventures (Ford and VW, Fiat and PSA), a strategy enabling both a reduction in production costs of standardised components and product diversification. A case in point is the Setúbal plant of Ford and VW with its initial capacity of 120,000 automobiles per year (see Ferrão and Vale, this volume). The competition in this market segment is rapidly intensifying, with the new Hordain plant of Fiat and PSA in northern France having an annual capacity of 130,000 vehicles made by 3,500 workers. At present, even Mercedes-Benz is refurbishing a van plant in Vitoria in Spain for production of a new model of multi-purpose vehicles. Within national production systems, it has been common practice, for many years, for governments to subsidize new investments in old industrialised locations, irrespective of the general philosophy of their economic policies. Thus, both France and Germany have tried to attract automobile investments to particular places via subsidies (Oberhauser 1987, 1990, Schamp in this volume). Old industrialised locations such as Hordain and Vitoria and new sites such as Setúbal profit increasingly from the emerging regional policy of the European Union Commission and from the willingness of governments lavishly to subsidize new investment. This, however, will bring to an end the above average 10 per cent annual profit rates enjoyed by Renault and Matra. While 117,000 multi-purpose vehicles models were sold in Europe in 1992, admittedly a year in crisis, most individual producers' capacities exceed this figure. As a consequence, low break-even points become increasingly vital, which in turn enhances the importance of seeking to reduce fixed costs.

In addition to experiencing the changes described above, western and southern Europe has formed the laboratory for experimenting with and learning about new production concepts. Earlier experiments with 'lean' production in the Nissan Sunderland plant have been echoed in new or renewed plants in the European production core, such as the 1986 BMW Regensburg plant, the VW plant at Emden or the Citroen/PSA plant in Rennes. The 1990s are witnessing a more radical reorganisation of production at some greenfield sites in southern Europe. New local production complexes are emerging in the early 1990s which apparently follow the Toyota model of spatial proximity of synchronised component assembly to flexible automobile assembly. The new Fiat plant at Melfi in the Mezzzogiorno producing the new Punto model serves as an exemplar. In a suppliers' industrial park adjacent to the Fiat plant, 16 suppliers for headlights, screws and bolts, bumpers, fuel tanks, doors, seats, sheet metal parts, exhaust systems, headlinings, window lifters, hand brakes, gaskets and cable harnesses are located (Mehl, 1993). In the so-called Fordland 13 component suppliers to the nearby Setúbal Ford/VW plant will be established (Ferrão and Vale, this volume). In 1993, the Parque Industrial de Premontaje PIPS opened its doors for 15 component suppliers to the new flexible Martorell plant of Seat, producing the new Ibiza model near Barcelona. It may be questioned, however, as to whether this will become *the* dominant spatial organisational model of the 1990s. Obviously, suppliers' parks only accommodate final component assembly while remaining dependent upon distant parts' suppliers. As a result, they only create a small number of jobs locally. What has been called 'deadline proximity' by Ferrão and Vale (this volume) can be achieved in other regions either without a particular industrial park, as has been shown at Eisenach and Mosel (see Schamp, this volume), or without any spatial proximity. The explanation evidently lies in the locational context or, as Mehl (1993) put it for the Melfi plant: 'the aspired close spatial relationship with suppliers can be seen as a tribute to particular Italian circumstances: strikes in the transport sector, bad road and rail linkages make more difficult the production-synchronous delivery from larger distances'. It may also be related to a learning process involving growing trust in inter-company relations and organisational concepts which increasingly renders spatial proximity unnecessary, as Lung and Mair (1993) have emphasised.

There are two comments to be made about the future significance of these local production complexes. First, such complexes may be thought of as constituting an archipelago in a hostile sea (Ferrão and Vale, this volume) or perhaps as a 'cathedral in the desert' (Grabher, 1992). In any case, such an interpretation challenges Porter's (1990) views as to the meaning of geographical concentration for global competitiveness since integration into the socio-economic environment is lacking. Second, new Japanese transplants such as Burnaston (Toyota) and Swindon (Honda), but also the Suzuki plant at Esztergom in Hungary, avoid devolving component assembly to suppliers and partly reintegrate pre-assembly in the automobile plant.

This brings the debate, finally, to the role of component suppliers in the emergence of automobile production outside the European core area. There has been a double movement of supplier firms to the European fringes. First, component suppliers followed automobile production. In fact, automobile producers from the European core (the American firms Ford and GM included) in particular preferred maintaining links to old supplier firms via sourcing from new component plants. In contrast, the Japanese looked more for domestic or European suppliers. As a result, domestic component producers in Spain have been largely crowded out by foreign suppliers (Lagendijk and van der Knaap, 1993) while German and French suppliers were able to invest in the United Kingdom. This, among other reasons, gave rise to the process of Europeanisation of component suppliers, analysed in more detail by Sadler and Amin (this volume).

In fact, Europeanisation of the automotive component industry towards western and southern Europe has taken place for reasons comparable to those which influenced the Europeanisation of automobile production in small-sized models: that is, a combination of newly emerging markets and availability of a qualified workforce for lower wage costs. In view of recent wage negotiations and sales crises, it is an open question as to how long these factors will continue to favour automobile production in these locations, particularly in southern Europe.

10.3 Central and Eastern Europe - the New 'Frontier' of European Automobile Production ?

When the European automobile producers faced a serious demand crisis at the end of the 1980s, central and eastern Europe appeared on the scene as promising newly emerging markets. Despite political instabilities and the economic break-down of old state socialist systems, there was a euphoric view among automobile producers as to the new opportunities on offer, leading to a 'gold rush' (Carroué, 1993) to these parts of Europe. Prospects were favourable: private automobile consumption was low compared to western Europe with, in 1991, 136, 187 and 208 cars per thousand inhabitants in Poland, Hungary and the Czech Republic, respectively; in addition, products and production processes in Poland and Czechoslovakia were obsolete and inefficient. Indeed, automobile registration exploded by 100 per cent whereas domestic production collapsed by 41 per cent in Poland in 1991. Most observers, however, failed to see that increased registrations mainly reflected imports of used cars from western Europe. Equally, they failed to appreciate the significance of the disintegration of the former economic system, with its distinctive state socialist spatial division of labour, either because of territorial changes via German re-unification or through the dissolution of the former Comecon. Consequently, national markets were re-established, differing from one another as a result of the differing routes and speeds of the national transition to capitalism. The legacies of socialist automobile production became nationally fragmented and were, finally, prepared in varying ways for integration into the emerging European automobile production system. The following paragraphs will mainly focus on Poland, Hungary and the Czech Republic as these areas are increasingly incorporated into the European economic system. There are several reasons for this focus. Automobile production in the CIS will be neglected. In fact, the current situation there is all too unclear. Furthermore, an economic region in its own right might conceivably emerge from the present disorder in the CIS. Post-unification automobile production in the former GDR has been discussed alongside production in the European core area (see Schamp, this volume). The decomposing Yugoslavia and other smaller countries of eastern Europe will not play a significant role in automobile production in the foreseeable future (for these see Sadler and Swain, 1993).

A combination of the inherited structures of automobile production and national policies of post-socialist governments will shape the character and form of their integration into a broader European production system. In Poland, two state firms existed: FSO in Warsaw producing an obsolete middle-sized model, the Polonez, based on a Fiat licence, and FSM in Bielsko-Biala where the small car 126 is still produced, also under licence from Fiat. These long-established contacts between the Polish government and Fiat company have been rejuvenated after 1989. Fiat subsequently installed production of a completely new model, the Cinquecento, in the Tychy works of FSM in 1991 and, one year later, acquired 90 per cent of FSM, changing its name into Fiat Poland SA. This is one of the largest foreign direct investment projects in Poland, involving almost USA$ one billion between 1993 and 1996. This small sized model, however, still is too expensive for most Polish clients. As a result, in 1993, 75 per cent of an annual production of 200,000 cars was exported to Italy, and a further 15 per cent to Germany, with only 10 per cent sold in eastern Europe. Obviously, in terms of purchasing power the Polish market still remains small. The Polish government has, nevertheless, chosen a highly protectionist policy, combined with an aspiration to diversify domestic suppliers. This clearly is a traditional import substituting policy, with all its well known effects. PSA started assembly of the Peugeot 405 model at Lublin, in 1993, and in the same year VW began to assemble a pickup in Poznan. Opel agreed to join FSO, assembling 10,000 Astras in 1994 while Fiat was allowed to assemble some Uno models in the Bielsko-Biala plants. Finally, VW was obliged to assemble Skoda models from July 1994 onwards in its Poznan plant as Polish customs duties increased the price of a Skoda model by 41 per cent, causing sales to collapse from 40,000 in 1991 to only 3,000 in 1993. Customs duties are even more than the 35 per cent Poland imposes on automobiles imported from the European Union - which, furthermore, the Polish government agreed to reduce slowly. Small assembly capacities for older models will certainly not attract large investments by foreign automobile producers and will only have minor impacts on technological modernisation. Additionally, a corollary of these agreements was to allow duty-free imports of component parts. For example, all engines and electronic devices and sheet metal parts for the Cinquecento are imported from Turin. As a result, beneficial impacts on the modernisation of the Polish component suppliers' industry will be strictly limited.

Integration of Czech automobile production into the western European production network followed a very different path. During socialist times, Czechoslovakia had invested into the well-known Skoda automobile plants at Mlada Boleslaw and Vrchlabi, in which automobiles had been built as early as 1905. The Favorit model, introduced in 1988, had even been exported to western Europe in large numbers. In 1990 the Czech government set in motion a bidding process among European automobile producers for a joint venture with Skoda. This, finally, was won by VW because of its acceptance by the Czech management and unions. Dankbaar (1994) has vividly reported on the different strategies of VW and Renault in this process. VW agreed to maintain the Skoda marque and to invest more than DM 7.0 billion. This, however, was reduced to only DM 3.7 billion in 1993 as VW's financial crisis was further aggravated by huge losses at Seat. This led to much negotiation with governments. As a result, plans to build an engine plant have been stopped; engine parts will be imported from Wolfsburg. Planned capacity increases will be reduced to an annual production of 400,000 cars (compared to one of 180,000 in 1989). Nevertheless, Skoda has been re-established as an independent marque exporting nearly 50 per cent of its production to western Europe. In part because of a rather low customs duty of 15 per cent on products from the European Union (compared to 19 per cent for imports from other countries), Skoda even made a small profit in 1993. Technical upgrading of Skoda models, however, which is so necessary if they are to become competitive on west European markets, was followed by higher sales prices. This resulted in increasing difficulties with sales both on the domestic market, in which the small-sized low-cost model Tavia from the Ukraine is winning a growing market share, and on foreign eastern European markets.

The legacy of a largely vertically integrated automobile production system in Czechoslovakia has been advantageous in other ways. Its dissolution has allowed the emergence of independent component suppliers and for a close integration into a more or less western European, or perhaps more accurately German, supplier system. Component suppliers to VW - and others - such as ITT, Rockwell-Golde, Bosch, T and N, VDO, Lucas, Leonische Drahtwerke and Benteler deliberately followed the assemblers to the Czech Republic via a series of joint ventures and take-overs. The component companies did so for two main reasons: first, in order to get access to the market for Skoda supplies; second, because parts production for the western European market proved to be profitable because of low wages, the relatively high productivity of the workforce and undervaluation of the Czech Crown.

Of all the formerly socialist countries, Hungary had the greatest experience with joint ventures. Some time before the other socialist states, Hungary had changed into a kind of semi-market economy with corresponding laws allowing and regulating private initiatives. This may have been one reason why capitalist automobile producers first pushed into Hungary after 1989. Other reasons include access to central and eastern European markets; low cost production for western European markets, although wages were lower in other former state socialist states - for example, they were twice as high as in Poland in 1991. Partly as a result of earlier contacts, some automobile producers invested quickly in greenfield plants. Opel developed a plant for engines in Szentgotthàrd, a small place near the Austrian border, where some Astra models are also assembled for the domestic market. Ford started a new plant for ignition coils at Székesféhvár, a medium-sized industrial town to the west of Budapest. Furthermore, Audi will produce components for its new generation of engines at Györ. The most interesting and radical experiment, however, involves the Japanese joint venture of Suzuki factories at Esztergom, located to the north of Budapest. This seems to run counter to the general hesitation that Japanese producers have clearly shown over investing in central and eastern Europe (Dankbaar 1994). Since 1992, Suzuki has produced an older small-sized model, the Swift, both for the domestic market and for export to the European Union. It has done so in a highly flexible manner and with a low degree of automation in a very small factory. In fact, it has a planned capacity of only 50,000 cars per year and a break-even point of approximately 30,000 which will be reached in 1994. Full integration from metal stamping to assembly is managed with some 700 employees. This can, however, only be achieved by reducing the product range to only two variants of one model, a hatchback for export and a saloon for the domestic market, without any variation in equipment - the production process may be flexible but the model choice looks suspiciously like a reversion to a supposedly bygone Fordist era. Engines and power train are imported from Japan, sheet metal and other supplies from west Europe, but some components are delivered from Hungarian suppliers. The problem of selecting efficient suppliers has been solved in a particular way. Suzuki asked its own suppliers in Japan to audit, select and train Hungarian firms. Although Suzuki has not yet achieved its agreed target of 60 per cent local content, it is nevertheless being allowed to export duty-free to the European Union as a result of a special customs agreement.

Relatively low investments, tax reductions granted in the early stages of the transition of the Hungarian economy and free access to western European markets have substantially insulated the automobile producers from risk. Nevertheless, enormous problems of deficient provision of physical and institutional infrastructure limit the success of the Hungarian 'model'. Consequently, greenfield sites remain 'islands in the modernisation process', with limited impact on labour markets and few regional spill-over effects (Czéfalvay et al., 1993). Some further examples of component suppliers relocating parts production from Germany to Hungary emphasise that Hungary is in competition with western and southern European countries such as Portugal and Ireland for such investment projects.

In view of the ongoing character of the transition process, it is too early to speculate about the future role of central and eastern Europe in an emerging European automotive production system. Alternative scenarios of isolationist national projects and complete integration into a western European system seem to be possible (see Bordenave and Lung, 1993). Much depends upon national governments' and European Union policies and, related to this, changes in domestic prosperity. It is, however, clear that central and eastern European countries will be integrated into a broader European system in very different ways. Provided with a marque of its own in the lower medium-sized market segment, a differentiated component suppliers industry and with VW, a global player with strong European targets, hovering in the background, it seems very likely that the Czech Republic will be integrated into the western European automobile core area more rapidly and easily than will the others. This is especially so as a combination of the revaluation of the Crown, low domestic inflation rates but increasing wage levels will transform the Czech Republic into a high wage country which will be forced to seek competitive advantages via the qualifications and skills of its workforce.

With regard to the other eastern European countries, it remains an open question as to whether past experiences elsewhere might be used in an analogous way to shape future prospects. Could, for instance, Poland follow the Spanish route of a rapid increase in car ownership and use combined with specialization in the production of small cars? In terms of population figures and, hence, market possibilities, the two initially seem comparable. In other important ways, however, they differ. Processes of modernisation and industrialisation had much greater impacts on Spain by the early 1970s than they have had on Poland by the early 1990s. Moreover, Spain had chosen a pathway of strong integration into western European markets, not one of protectionism. Could Hungary use its initial lead in the processes of transition to a market economy and follow the strategy that Austria had previously successfully implemented? During the 1980s, the Austrian government increasingly forced automobile importers either to invest in Austria (which led to the GM engine plant in Aspern near Vienna) or to buy parts and components from Austrian suppliers (which led to the emergence of an efficient supplier sector). Subsequently, this supply base was influential in Chrysler's decision to locate automobile assembly in its Graz plants. In the present context, however, it seems unlikely that the Hungarian government will be in a sufficiently strong bargaining position successfully to build up a qualified suppliers' sector. This will bring the country nearer to Ireland than to Portugal as a low wage and weak currency location for subcontracting.

10.4 Resistance to the Erosion of the European Automobile Core

The re-emergence of automobile production in the United Kingdom through Japanese transplants (Hudson, this volume) and the relocation of the centre of gravity of production of medium-sized volume models from northern Italy to the southern triangle of Melfi, Pomigliano and Cassino (Conti and Enrietti, this volume), together with the rise in Spanish automobile production, clearly suggest a long-run relocation process from the traditional western European core areas to European semi-peripheries. The current crisis and the reduction of the workforce in the core reinforces that suspicion. It is, however, not verified by production figures during the last decade. Production rose considerably in Germany, France and, particularly, Belgium during the 1980s, countries in which principally both high volume medium-sized models and luxury automobiles are produced (Fig. 10.2). What do these figures mean in the 1990s, as the market shares of medium-sized saloon automobiles are declining as demand increases for 'pleasure cars' such as off-road vehicles, convertibles, multi-purpose vehicles and the like and as luxury car production increasingly comes under pressure from Japanese competition? In response to these tendencies, there is evidence of various forms of resistance towards this decentralisation of automobile production to the European fringes in the responses of governments, trade unions and works councils and even enterprises. In general, it involves the search for new methods of production that will be commercially viable in the high wage environment of the European core. This encompasses the implementation of new production concepts at old sites, a concentration on core capabilities, the deregulation and reregulation of labour markets, and a reorganisation of the spatial division of labour combined with increasing concentration of production as companies forge still more

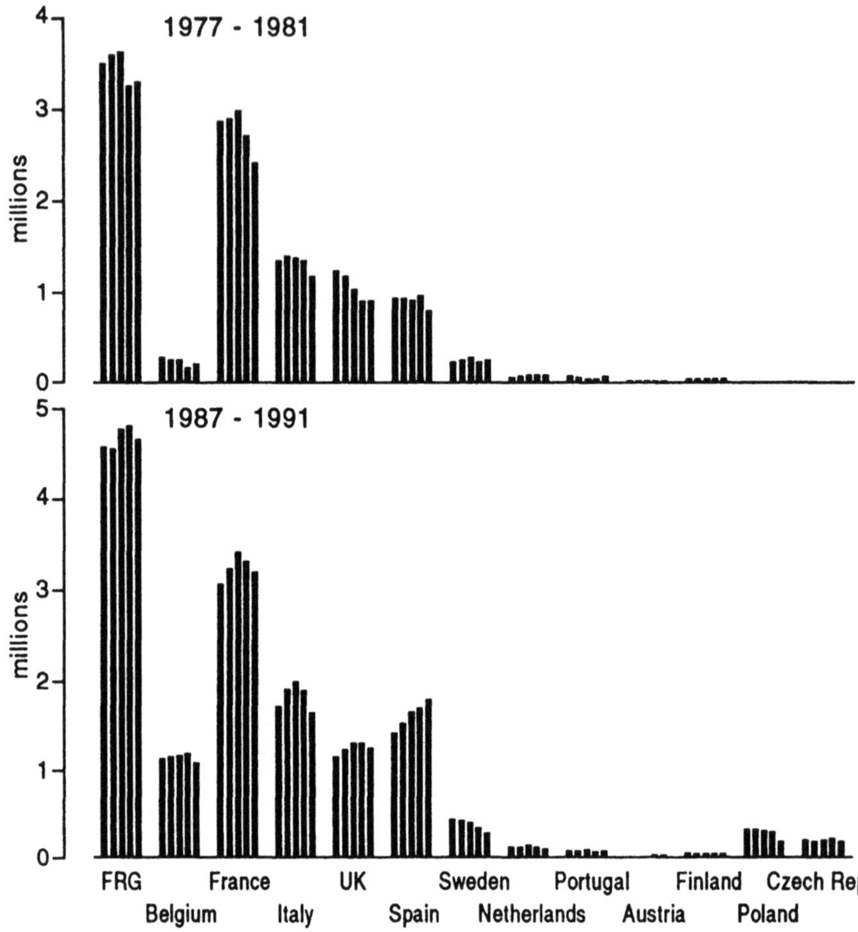

Fig. 10.2 Production of automobiles in major European countries, 1977-81 and 1987-91.

strategic alliances.

There are numerous examples throughout this book of such learning processes in the European core areas as various agents seek to adjust to and cope with new conditions for automobile production. New production concepts, variously designated as Japanisation, Toyotism, lean production or even hybridisation, are now common currency among the automobile producers. Implementation of such concepts, however, varies according to a particular firm's philosophy and the specific societal characteristics of the home country. The consequences of these individual corporate responses are increasing disintegration of national production systems and integration into European production networks. It has to be emphasised, however, that the strategies of individual automobile producers share one crucial common characteristic: that is, an upgrading of the functions of the European core. One way of doing this is to re-define the core capabilities of the firm. Varying examples of recent decisions to allocate engine production to German sites illustrate the point: Opel has located production of diesel engines in Kaiserslautern and lean burn energy efficient engines in Bochum; Ford has allocated production of the new large engines, mainly for export to the United States of America, to Köln; KHD has also put its production of diesel engines in Köln; VW has allocated the production of engines to be fitted in Skoda models to Wolfsburg; and last but not least Mercedes' decisions about investment at Bad Cannstadt emphasise that engine production is increasingly becoming the central capability of the European core area. Shifting a part of R&D towards first tier systems suppliers, as part of long-term collaborative agreements, is one way in which automobile producers increasingly seem to safeguard technical competence in engine production and development.

There is, furthermore, a simultaneous process of strengthening competences in R&D in the European core, both by automobile producers and major component suppliers. Automobile producers considerably increased their R&D expenditures in recent years, counter to the trend to allocate more R&D to systems suppliers. Opel and Ford, for instance, dedicated more than 5 per cent of annual turnover to R&D in 1993, a level that they had not reached before. Automobile production is clearly seeking to regain its status as a high tech sector. Bordenave and Lung (1993) give some examples of huge R&D centres : for instance, the world R&D centre of GM in Rüsselsheim (with responsibilities for everywhere except North America) has more than 7,000 employees, those of Renault at Guyancourt in Saint Quentin, west of Paris (formerly, until 1991, in Rueil in Paris) and BMW in Munich each have more than 6,000 employees, while those of Ford in Köln and Mercedes-Benz in Sindelfingen both have more than 4,000 employees. There is further evidence of processes of concentration of R&D in Italy and France (Conti and Enrietti, Savary, this volume). In general, key functions in decision-making and R&D mainly remain in the European core, reinforcing a hierarchical order between the centres and branch plants in the European fringes. This holds true for large component suppliers as well. It is in this realm that the United Kingdom lost a good part of its competitiveness when Ford relocated R&D functions from Dunton to Köln and Nissan located its European design centre in Brussels. The takeover of Rover by BMW might have comparable effects in the long run. In 1994 it was announced that much of development responsibility for the new Mazda 323 model had been given to its European design centre near Frankfurt. This process of further spatial concentration of R&D also clearly illustrates the struggle for R&D centres between locations within the same company. For example, both Seat (Spain) and Skoda (Czech Republic) are competing with VW's core plant at Wolfsburg for R&D functions.

At the same time, pursuit of cost reduction in the core area remains an important issue. This is partly achieved by introducing new concepts in procurement and distribution logistics, as exemplified with the GM Consolidation Centre at Bochum for central procurement of European sites in Belgium, the United Kingdom and Portugal, or the recently announced distribution centre for all Ford models produced in Germany, Belgium and the United Kingdom at Neuß. This latter mainly will use rail and ship as modes of transportation. More attention has been paid to cutting labour costs, however, either by reducing voluntarily paid incidental costs, redundancies or re-regulation of work. Recent attempts to increase the flexibility of the labour in automobile plants without increasing costs have led, for example, to more than 200 different agreements with the works council of BMW seeking to enhance the competitiveness of traditional production plants (see Schamp, this volume). In general, new methods of work organisation reduce labour costs but safeguard existing skill requirements and knowledge - and even enhance them - in the European automobile production core areas. Industrial relations based on capital:labour cooperation and mutual trust rather than adversarial competition became a major locational factor in favour of the European core area (Hudson, this volume). In consequence, production of new generation small sized models is increasingly being relocated to the core area. For example, the new Polo is being produced at Wolfsburg (as well as at Pamplona) and production of the A model of Mercedes-Benz is planned to commence in 1997 at Rastatt. The growing importance of non-confrontational industrial relations will become of particular relevance when the now young labour forces in highly flexible plants such as Sunderland, Eisenach or Mosel grow older and, as a result, the productivity of such plants will decrease. The long-run maintenance of flexibility can only result from a broader societal acceptance of such forms of work organisation. If this is lacking, a more pronounced relocation of production to the European fringes becomes necessary, as has obviously been the case in Italy (Conti and Enrietti, this volume; Mehl, 1993).

Strategic reorganisation of automobile producers on a global scale also contributes to a strengthening of the upper echelons of the corporate hierarchy in the European core areas. Ford recently announced a major shift in its organisational structure, implementing a kind of global organisation and allocating global responsibilities to different subcentres. Köln and Dunton will jointly be responsible for all small and medium-sized models, including the Fiesta, Escort, Mondeo and Probe. Some commentators quickly recalled Ford's disastrous earlier world car experiment. This latest experiment may end, however, in a mastering of the problems of global coordination and control and undoubtedly offers possibilities for standardisation and the realisation of scale economies in production and development. Currently, the hitherto national - if not regional - specialists BMW and Mercedes-Benz are taking steps towards globalisation. BMW is building up an assembly plant for a daily production of 300 to 400 roadsters, both for the North American market and export to Europe, at Spartanburg in South Carolina. From 1995, four jumbo jets will bring engines from Germany every day. Mercedes-Benz announced a new assembly plant for off-road vehicles in Tuscaloosa in Alabama which will lead to the relocation of production of off-road vehicles from the Steyr plant in Austria to the United States of America. Whereas during the 1980s volume producers acquired specialists - for instance GM, Fiat and Ford acquired Saab, Alfa Romeo and Jaguar, respectively - in early 1994 a wealthy specialist (BMW) took over the majority of a smaller volume producer (Rover). This merger opened new market segments to BMW, such as small automobiles (Mini, Rover 100) and off-road vehicles (Landrover, Range Rover), broadening the range of BMW's products. It also enabled BMW to deliver one of its core products, a diesel engine, to Rover - remember that BMW means Bayerische Motoren-Werke, Bavarian Engine Factories - and to achieve increasing economies of scale through standardisation of, for example, the body platforms for the BMW 5 and Rover 800. Speculations about further mergers among European automobile producers remain rife but recent joint ventures in the multi-purpose vehicle's sector appear very tempting, such as that between Fiat and PSA. Or will there be a Renault-Fiat deal, as Savary (this volume) suggests? In any case, future mergers among the European automobile producers still seem further to reinforce a spatial concentration of decision-making power in the traditional core area.

Societal resistance to the decentralisation of automobile assembly to the European fringes is strong, for three reasons: first, the sheer size of the plants involved; second, because of effective forms of labour resistance and political pressure; and third, as a result of organisational innovations in spheres such as work organisation, design and logistics. In contrast, out-migration of automobile suppliers to low wage countries is a much larger, though publicly less acknowledged, process. Concentration of component provision in a few system suppliers - called 'first tier' according to the Japanese experiences - not only means an increasing division of labour between component suppliers but an increased division of labour between the different plants of those system suppliers as well. In consequence, there is a tremendous restructuring of the spatial division of labour in the supplier sector, leading to the relocation of parts production to the European fringes, or even further in the absence of great societal resistance (Sadler, 1994). Certainly, core capabilities, decision-making and R&D still remain in the European core areas, as is the case with automobile assembly. Additionally, a small number of pre-assembling satellite plants have had to be established in close spatial proximity to the automobile assembly plants, even in the European core. This trend is currently exemplified by developments around the Ford plant in Genk and the VW plant at Wolfsburg. Given the diversity of the sector, a size structure still dominated by medium-sized firms, and the relative weakness of both firms and unions, it is very likely that the relocation process will continue.

10.5 Towards a New Map of Automobile Production in Europe?

The central thesis of this book is that different national production systems in the European automobile industry increasingly are being called into question and that, as a result, some kind of a new European production system may emerge. It is, however, far too early to draw a clear map of this possible future production system since many contradictory processes are currently at work. This makes prediction a rather difficult and dangerous task. Bordenave and Lung (1993) recently discussed in more depth different ideal scenarios for the spatial reconfiguration of automobile assembly in Europe. We will not add to this very useful discussion. Instead, we will try to identify some key characteristics in the contemporary reorganisation of automobile production systems. In doing so, we will, consciously, focus on the observable manifestations of spatial change.

Fig. 10.3 and Table 10.1 reveal that automobile assembly is still principally located in the traditional European core areas, ranging from southern England, Belgium and northern France via western Germany to northern Italy. It depicts the spatial pattern of automobile assembly and component suppliers in the 'European core' recognised by Jones (1993). Although the market shares of their products are decreasing, these plants will remain the strategic centre of automobile production. High and continuing investments will strengthen their position. There will more likely be a kind of hidden deindustrialisation, in two senses: with further increases in productivity, employment will decline; there will also probably be a hidden reduction in production capacities through model changes. Core products, core capabilities and upgrading strategies, though, will foster the role of the traditional core areas as the top of the hierarchical pyramid of a European production system.

The relocation of the assembly of special models and small-sized models to more peripheral sites in Europe, which is clearly shown by Figure 10.3, may not be simply the outcome of an old Fordist model of spatial organisation in the automobile production system. In fact, a horizontal specialization is emerging among European assembly plants as new models and new production concepts are initially introduced in some more peripheral parts of Europe. This changed geography of assembly is juxtaposed to the domination of financial control and R&D from the centre. Automobile assembly in the European fringes may be more vulnerable than usually is recognised (see Hudson, 1995). In some plants the more capital intensive production stages, such as stamping, are absent. Public sector subsidisation of investments in the European fringes further reduces capital costs. Assembly is also concentrated on niche products or on the small-sized model sector, in which there is increasing competition between southern (Opel, Ford, Seat) and eastern (Fiat, Skoda) European plants. Finally, given the still diversified monetary landscape in Europe, currency exchange

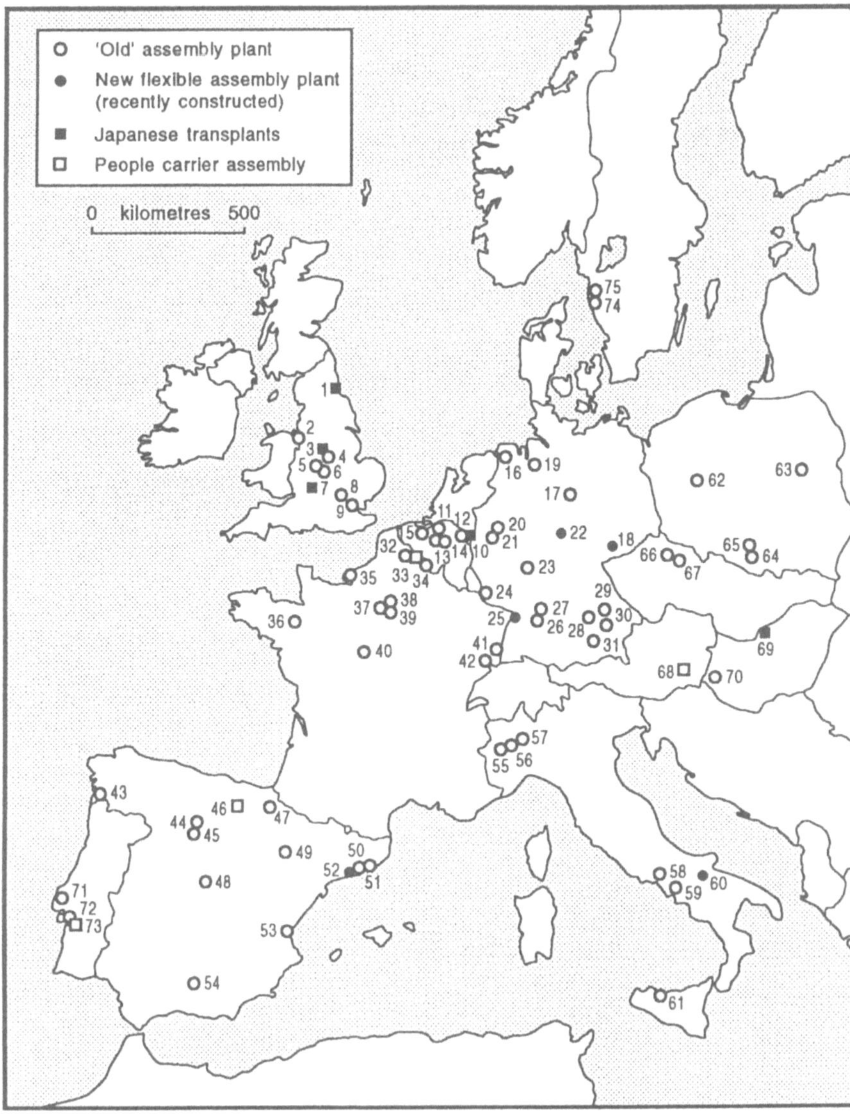

Fig. 10.3 Automobile assembly plants of major importance in Europe, 1994 to 1996. (for key, see Table 10.1)

rate alterations will further affect competitiveness. The net effect of these various interacting variables could well be to produce rapid changes in the geography of production within the European peripheries.

It may perhaps be more of a surprise to some to realise that in the component supply sector the old 'Fordist' model of spatial organisation of production will remain viable throughout the 1990s. Indeed, it may possibly become more

Table 10.1 Location of major automobile assembly plants in Europe and key to Fig. 10.3

Number	Place	Firm	Type
United Kingdom			
1	Sunderland	Nissan	C
2	Halewood	Ford	A
3	Burnaston	Toyota	C
4	Ryton	PSA	A
5	Longbridge	Rover	A
6	Cowley	Rover	A
7	Swindon	Honda	C
8	Luton	GM	A
9	Dagenham	Ford	A
Netherlands			
10	Born	Volvo/Mitsubishi	C
Belgium			
11	Antwerp	GM	A
12	Genk	Ford	A
13	Brussels	VW	A
14	Brussels	Renault	A
15	Ghent	Volvo	A
Germany			
16	Emden	VW	A
17	Wolfsburg	VW	A
18	Mosel/Zwickau	VW	B
19	Bremen	Mercedes-Benz	A
20	Bochum	Opel/GM	A
21	Köln	Ford	A
22	Eisenach	Opel/GM	B
23	Rüsselheim/Frankfurt	Opel/GM	A
24	Saarlouis	Ford	A
25	Rastatt	Mercedes-Benz	A
26	Sindelfingen	Mercedes-Benz	A
27	Neckarsulm	Audi	A
28	Ingolstadt	Audi	A
29	Regensburg	BMW	A
30	Dingolfing	BMW	A
31	München	BMW	A
France			
32	Douai	Renault	A
33	Valenciennes	Fiat/PSA	D
34	Maubeuge	Renault	A
35	Sandouville	Renault	A
36	Rennes	PSA	A
37	Flins/Paris	Renault	A
38	Paris/Poissy	PSA	A

39	Paris / Aulnay	PSA	A
40	Romorantin	PSA	A
41	Mulhouse	PSA	A
42	Sochaux	PSA	A
Spain			
43	Vigo	PSA	A
44	Palencia	Renault	A
45	Valladolid	Renault	A
46	Vitoria	Mercedes-Benz	D
47	Pamplona	VW	A
48	Madrid	PSA	A
49	Zaragoza	Opel / GM	A
50	Barcelona	Nissan	A
51	Barcelona	Seat	A
52	Martorell / Barcelona	Seat	B
53	Valencia	Ford	A
54	Jaen	Suzuki	A
Italy			
55	Turin / Rivalta	Fiat	A
56	Turin / Mirafiori	Fiat	A
57	Arese	Fiat	A
58	Cassino	Fiat	A
59	Pomigliano	Fiat	A
60	Melfi	Fiat	B
61	Termini Imerese	Fiat	A
Poland			
62	Poznan	VW	A
63	Warzaw	Opel	A
64	Bielsko Biala	Fiat	A
65	Tychy	Fiat	A
Czech Republic			
66	Mlada Boleslaw	Skoda	A
67	Vrchlabi	Skoda	A
Austria			
68	Graz	Chrysler	D
Hungary			
69	Esztergom	Suzuki	C
70	Szentgotthárd	GM	A
Portugal			
71	Azambuja	GM	A
72	Setubal	Renault	A
73	Setubal	Ford / VW	D
Sweden			
74	Göteborg	Volvo	A
75	Trollhättan	Saab / GM	A

prevalent, reinforced by the new competition from eastern Europe. This low wage competition will be largely underpinned by institutional arrangements, regulating factors such as annual hours of work, subsidies and currency exchange rates.

Finally, we may speculate about a possible double movement in the spatial reorganisation of automobile production systems in Europe. On the one hand, traditional core-periphery structures will continue to exist, explicable by older concepts in economic geography such as the spatial division of labour, or more broadly and in a more recent approach, characterised as the Fordist model of spatial organisation (Lung, 1992; Bordenave and Lung, 1993). These will be overlain by an increasing specialization and integration of automobile assembly. This, however, depends upon the possibilities for continuing strategies of upgrading quality in model development. In turn, this depends upon a continuing increase in purchasing power and stability in consumer preferences in the main European markets. As has been shown with the rather unsuccessful S model of Mercedes-Benz, this can not necessarily be taken for granted.

While labour has taken a certain role in the reorganisation of the automobile production system, the chapters of this book have not primarily focused on the social costs and gains stemming from such changes in the volume and types of available employment. Such issues are nonetheless important ones. On the one hand, most authors would reject the idea of a complete deindustrialisation leading to the cessation of automobile production in the traditional European core areas (for example, Lung, 1992; Bordenave and Lung, 1993). Nevertheless, there is a generalised awareness of an increasing 'hollowing-out' of employment in those regions. While automobile production increased considerably during the 1980s in Europe, employment declined in some countries, particularly in Italy, France and the United Kingdom. At the same time, increased employment in Spain, Belgium, the Netherlands and Germany clearly underlined the absence of increased productivity, something which will probably be redressed during the 1990s (Belzer and Dankbaar, 1994). It is in these countries, as well as in eastern European countries experiencing the particular problems of transition, that employment in the automobile production system will be drastically reduced. In fact, this is already happening during 1994 in countries such as Germany, both in assembly and more silently in the components sector. Since 1991, employment in Germany, which was by far the largest national automobile production system in Europe (44 per cent of total European employment) has already been reduced by 115,000 or 15 per cent. It is forecast to decrease by as much as a further 200,000 in the coming years. This process of employment reduction is not, however, restricted to the automobile production systems in the European core but extends to other manufacturing sectors as well. Whether a relocation of manufacturing within Europe will affect the export markets for automobiles produced in the European fringes because of reduced purchasing power in the European core, especially given the economic importance of the sector for most western European national economies, remains an open question. The way in which it will be resolved will largely depend upon societal and political responses in the core economies.

Other environmental and social costs are generated by the excessive use of private cars in the major European markets. New institutional arrangements have been (and are being) invented in the core areas to prevent the collapse of circulation because of an overwhelming mass motorization. New regulatory frameworks are being sought to contain the ecological and health impacts of environmental pollution and the personal and social costs arising from traffic accidents and time wasted because of congestion. These are also aimed at safeguarding demand and markets for private automobiles, however. In this way, such regulations simultaneously contribute to the survival of the current trajectory in traffic technologies. At present, in peripheral regions of Europe, particularly in eastern Europe, institutions and infrastructures are being established to facilitate a rapid convergence in mass motorisation between those regions and western Europe. This will further gridlock traffic technologies into existing trends and rule out the exploration of other possibilities of managing mobility. It seems as if the possibility of a technological revolution supplanting the use of privately-owned automobiles, powered by fossil fuel engines, remains a very distant one. Equally, so too does the possibility of a really radical shift in the geography of the automobile production system.

Bibliography

ACAP (1992): O Comércio e a Indústria Automóvel em Portugal, Lisbon, Associação do Comércio Automóvel de Portugal

ACP (1992): Revista do Automóvel Club de Portugal, No. 12, December, 1992

Afia (1992): Estudo sobre Tecnologias e Capacidades Laboratoriais no Fabrico de Componentes para a Indústria Automóvel, Oporto, Associação de Fabricantes para a Indústria Automóvel

Amin, A. (1985): Restructuring in Fiat and the decentralisation of production in southern Italy. In: Hudson, R., and Lewis, J., (eds.), Uneven Development in Southern Europe. Methuen, London, pp. 155-91

Amin, A. (1986): Restructuring in Fiat and Decentralisation of Production in Southern Italy, Mimeo. Paper presented at the London School of Economics, Autumn

Amin, A. and Smith, I. (1990): Decline and restructuring in the UK motor vehicle components industry. Scottish Journal of Political Economy 37, pp. 209-40

Amin, A. and Smith, I. (1991): Vertical integration or disintegration? The case of the UK car parts industry. In: Law C. (ed), Restructuring the global automobile industry : national and regional impacts. Routledge, London, pp. 169-99

Amin, A. Thrift, N. (eds.) (1994): Globalisation, institutions and regional development in Europe. Oxford University Press, Oxford

Appadurai, A. (1990): Disjuncture and difference in the global cultural economy. In: Theory, Culture and Society, 7 : pp. 295-310

Auer, P. (ed) (1991): Workforce adjustment patterns in four countries : Experiences in the steel and automobile industry in France, Germany, Sweden and the United Kingdom. WZB (Wissenschaftszentrum Berlin für Sozialforschung). Discussion Papers FS I 91-4, Berlin

Bailly, A. Maillat, D. (1988): Le secteur tertiaire en question. Economica, Paris

Balliano, P. (1986): Crisi e ristrutturazione del settore automobilistico negli anni '80. In: Michelsons A. (ed), Tre incognite per lo sviluppo. Franco Angeli, Milano

Banville, E. de. and Chanaron, J.J. (1991): Vers un système automobile européen. Economica, Paris

Banville, E. de., Chanaron, J.J. (1993): Les fournisseurs d'équipements automobiles. In: Sachweld, F. (ed) L'Europe et la globalisation. Acquisitions et accords dans l'industrie. Masson. Paris, pp. 113-161

Beccatini, G. (1991): The industrial district as a creative milieu. In: Benko, G. and Dunford, M. (eds), Industrial change and regional development: the transformation of new industrial spaces. Belhaven, London, pp. 102-114

Becchi Collida' A. and Negrelli, S. (1986): La transizione nell'industria e nelle relazioni industriali. L'auto e il caso Fiat. Angeli, Milano

Berg, H. (1984): Automobilindustrie. In: Oberende, P. (ed), Marktstruktur und Wettbewerb in der Bundesrepublik Deutschland. Branchenstudien zur deutschen Volkswirtschaft, Vahlen, München, pp. 169-215

Berggren, C. (1990): Det nya bilarbetet. Konkurrensen mellan olika produktionskoncept i svensk bilindustri 1970-1990. Lund: Arkiv avhandlingsserie

Bertram, H. (1994): Neue Produktionskonzepte und Wandel der Standortstruktur. (forthcoming)

Best, M.H. (1990): The new competition. Institutions of industrial restructuring. Polity Press, Cambridge

Beynon, H. (1973): Working for Ford. Penguin, Harmondsworth

Beynon, H. (1986): Nissan : the Japanese Car Industry and the Washington Connection. Unpublished manuscript, University of Durham

Bhaskar, K. (1990): The effect of different state aid measures on intra-Community competition, exemplified by the case of the automotive industry. Commission of the European Communities

Bloomfield, G.T. (1978): The world automotive industry: problems in modern geography. Newton Abbot

Bloomfield, G.T. (1991): The world automotive industry in transition. In: Law, C. M. (ed): Restructuring the global automobile industry. National and regional impacts, Routledge, London, pp. 19-60

Bochum, U. and Meißner, H.R. (1990): Im Tiefflug durch die Handelsschranken. Standortstrategien der großen Automobilkonzerne. In: Muster, M, and Richter, U, (eds): Mit Vollgas in den Stau. VSA-Verlag, Hamburg, pp. 13-37

Böhmer, R. (1992): Rund um die Uhr. Mit neuen Konzepten wächst der Druck auf west deutsche Standorte. Wirtschaftswoche 46(11), 6 March, pp. 190-2

Bonazzi Giuseppe (1993): II tubo di cristallo. Il Mulino, Bologna

Böndel, B. (1992): Schritt voraus. Sensoren, Informationssysteme und neue Werkstoffe verbessern die Wirtschaftlichkeit kommender Fahrzeuggenerationen. Wirtschaftswoche 46(33), 7 August, pp. 50-56

Borlenghi, E. and Dematteis, G. (1982): Localizzazione industriale e politica territoriale nell'area torinese. In: Torino - Detroit: due citta' a confronto. Citta' di Torino, Torino, pp. 109-131

Boston Consulting Group/PRS (1990): The EC automotive components sector in the context of the single market. London

Boston Consulting Group and PRS Consulting International (1991): The Competitive Challenge Facing the European Automotive Components Industry, Devonshire House. London WIX 5FH

Boyer, R. Durand, J. P. (1993): L'aprés-fordisme. Syros, Paris

Butera, F. (1991): Il castello e la rete. Franco Angeli, Milano

Cappellin, R. (1991): Patterns and Policies of Regional Economic Development and Cohesion Among the Regions of the European Community. Paper presented at the workshop The Restructuring of the Polish Social and Economic System: a Regional Approach, Karlsruhe

Castronovo, V., (1971): Giovanni Agnelli. U.T.E.T., Torino

C.E.C. (1992): A Indústria Europia dos Veículos a Motor : Situaçåo, Problemas em Queståo e Propostas de Acçă. COM (92) 166, Brussels, Commision of the European Communities

Cecchini, P. (1988): The European challenge: 1992 - the benefits of a single market. Wildwood, Aldershot

Central Policy Review Staff (1975): The Future of the British Car Industry. HMSO, London

Cerutti, G., and Rieser, V. (1991): Fiat: qualita' totale e fabbrica integrata. Ediesse, Roma

Chanaron, R. (1988): Productivity, vertical integration and competitiveness - some methodological reflections and empirical evidence. In: Dankbaar, B., Jürgens, U, Malsch, Th., (eds), Die Zukunft der Arbeit in der Automobilindustrie. Edition Sigma, Berlin, pp. 283-298

Collins, T. M. and Doorley, L. (1992): Les alliances stratéques. Inter Editions, Paris

Comito, V. (1982): La Fiat tra crisi e ristrutturazione. Editori Riuniti, Roma

Commission of the European Communities (1989a): A single Community motor vehicle market. SEC(89) 2118, Brussels

Commission of the European Communities (1989b): The future of the motor industry. SEC(89) 2275. Brussels

Commission of the European Communities (1992a): Intra-EC car price differential report. Brussels

Commission of the European Communities (1992b): The European motor vehicle industry : situation, issues at stake and proposals for action. COM(92) 166, Brussels

Conti, S. (1986): Il caso di Torino. La fase urbana e metropolitana dell a localizzazione industriale. In: Conti S., Lusso G. (eds), Aree e problemi di una regione in transizione. Patron Bologna

Coriat, B. (1990): L'atelier et le robot. Essai sur le fordisme et la production de masse à l'age de l'électronique. Christian Bourgois, Paris

Coriat, B. (1991): Technical flexibility and mass production. In: Benko, G. and Dunford, M. (eds.), Industrial change and Regional Development. Belhaven, London, pp. 134-58

County of Norrbotten (1989): Biltestverksamhet: Gasen i botten föîr Norrbotten. Länsstyrelsen i Norrbottens län. Rapportserie Nr. 1/1989, Luleą

Crowther, S. and Garrahan, P. (1988): Invitation to Sunderland : corporate power and the local economy. Industrial Relations Journal, 19, pp. 51-9

Dankbaar, B. (1990): International competition and national institutions : the case of the automobile industry. In: Freeman, C, and Soete, C. (eds), New explorations in the economics of technological change. Pinter, London, pp. 157-73

Davenport, E., Benington, J. and Geddes, M. (1990): The future of European motor industry regions : new local authority responses to industrial restructuring. Local Economy 5(2) pp. 129-46

Department of Industry (1975): British Leyland : the next decade. HMSO, London

Derbyshire County Council (1990): Toyota in Derbyshire - An Impact Study. County Offices, Matlock, DE4 3AG

Dicken, P. (1992a): Europe 1992 and strategic change in the international automobile industry. Environment and Planning A, 24, pp. 11-31

Dicken, P. (1992b): Global shift : The internationalization of economic activity, second edition. Paul Chapman Publishing, London. Guildford Publications, New York

Dicken, P., Thrift, N. (1992): The organization of production and the production of organization : why business enterprises matter in the study of geographical industrialization. Transactions Institute of British Geographers NS 17, pp. 270-291

Diez, W. (1988): Vertikale Diffusion: Zur Ausbreitung technischer Neuerungen auf dem deutschen Automobilmarkt. ifo-Schnelldienst 41,29, pp. 20-26

Diez, W. (1993): Deutsche Hersteller setzten auf 'mehr Auto für mehr Geld' und überließen neue Marktnischen den Japanern. Handelsblatt 86, 20

Dolata, U. (1990): Modernisierung und Umprofilierung der Konzernstrukturen in der Automobilindustrie. WSI Mitteilungen (1), pp. 37-51

Doleschal, R. (1991): Daten und Trends der bundesdeutschen Automobil-Zulieferindustrie. In: Mendius. H.G. and Wendeling-Schröder, U. (eds), Zulieferer im Netz. Neustrukturierung der Logistik am Beispiel der Automobilzulieferung, Bund-Verlag Köln, pp. 35-60

Dunford, M. and Benko, G. (1991): Neo-Fordism or post-Fordism? Some conclusions and further remarks. In: Benko, G, and Dunford, M. (eds), Industrial change and regional development: the transformation of new industrial spaces. Belhaven, London, pp. 286-305

Economist Intelligence Unit (1992a): Japan's Motor Industry - A Perspective on the Future. Special Report No. 2200, EIU, 40 Duke Street, London WIA 1DW

Economist Intelligence Unit (1992b): The European Automotive Components Industry: A review of the Industry and 80 Leading Companies. EIU, 40 Duke Street, London WIA IDW

Economist Intelligence Unit (1992c): The UK Automotive Components Industry - 1992 Edition. Special report No. 2158. EIU, 40 Duke Street, London WIA LDW

Ellegård, K. (1983): Människa - produktion. Tidsbilder av ett produktionssystem. Meddelanden från Göteborgs universitets geografiska institutioner. Serie B Nr 72. Göteborg

Elliot, D. and Gray, P. (1991): Policy implications of trends and changes in the vehicle and component industries : the case of the West Midlands. In: Law, C. (ed), Restructuring the global automobile industry : national and regional impacts. Routledge, London, pp. 251-62

Enrietti, A. (1987): La dinamica dell'integrazione verticale alla Fiat Auto Spa (1979-1986). In: Economia e politica industriale, no 55

Enrietti, A., Follis, M. and Fornengo, G. (1988): Innovazione tecnologica e potere di mercato nei rapporti tra imprese: l'industria dei componenti per auto. Angeli, Milano

Enrietti, A. and Fornengo, G. (1989): Il gruppo Fiat. La Nuova Italia Scientifica, Roma

Eriksen, K. (1990): Das Marktverhalten der Automobilhersteller auf den KFZ-Teile-Märkten der Bundesrepublik Deutschland. Wirtschaftspolitische Studien 82. Göttingen: Vandenhoeck & Ruprecht.

Euromonitor Reports (1991): Euromonitor Financial Yearbook, 1991. 105/6 New Bond Street, London, WIY 9GC

European Communities Committee (1990): A single market for cars. House of Lords paper 76, session 1989-90

European Parliament (1991): Report drawn up on behalf of the Committee on Economic and Monetary Affairs and Industrial Policy on the European car industry, Rapporteur : Carole Tongue. PE 140/91, Brussels and Strasbourg

Fairbrother, P. and Waddington, J. (1990): The politics of trade unionism : evidence, policy and theory. Capital and Class, 41, pp. 15-56

Fennetau, H. (1985): La structuration d'une filiere et l'organisation des relations entre les firmes. Le cas des industries lie'es a' l'automobile, in A.D.E.F.I., L'analyse de filiere. Economica, Paris

Fix, D. (1992): Im Grunde machtlos. Moderne Logistik und rasch wechselnde Produktvarianten machen Unternehmen bei Streiks verwundbarer denn je. Wirtschaftswoche 46 (12), 13 March, 58

Gabert, G. (1964): Turin, ville industrielle. P.U.F., Paris

Gabetti, R. (1977): Architettura, Industria, Piemonte negli ultimi cinquant'anni. Cassa di Risparmio di Torino, Torino

Gaebe, W. (1991): Internationalization of the Automobile Industry in Japan and in Europe (Volkswagen and BMW as examples) (mimeo)

Gamble, A. (1988): The Free Economy and the Strong State: the Politics of Thatcherism. Macmillan, London

Garrahan, S. (1986): Nissan in the north east of England. Capital and Class, 27, pp. 5-13

Garrahan, P. and Stewart, P. (1992): The Nissan Enigma : Flexibility at Work in a Local Economy. Mansell, London

Gepie (1990): Indústria e Energia. Evolução Recente e Perspectivas - 1990-1991. Lisbon, Ministry of Industry and Energy

Gertler, M.S. (1988): The limits to flexibility: comments on the post-Fordist vision of production and its geography. In: Transactions of the Institute of British Geographers, New Series 13, pp. 419-32

Giunepero, L. C. and O'Neal, C. (1988): Obstacles to JIT Procurement. Industrial Marketing Management, Vol. 17, pp. 35-41

Grabher, G. (1991): Eastern Conquista: The 'Truncated Industrialization'of East European Regions by Larger West European Corporations (mimeo)

Grabher, G. (ed) (1993): The embedded firm. On socioeconomics of industrial networks. Routledge, London and New York

Gupta, Y.P. and Heragu, S. (1991): Implications of implementing just-in-time systems. Technovation 11, 3, pp. 143-62

Helper, S. (1993): An exit-voice analysis of supplier relations. The case of the US automobile industry. In: Grabher G (ed), The embedded firm. On socioeconomics of industrial networks. Routledge, London and New York, pp. 141-160

Hill, R. C. (1989): Comparing transnational production systems : the automobile industry in the USA and Japan. International Journal of Urban and Regional Research 13, pp. 462-480

Hoffman K. and Kaplinsky, R. (1988): Driving force : the global restructuring of technology, labour and investment in the automobile and components industry. Westview Press, Boulder Colorado

House of Lords Select Committee on the European Communities (1990): A Single Market for Cars. House of Lords Paper, No. 76, HMSO, London

Hudson, R. (1989): Wrecking a Region. Pion, London

Hudson, R. (1991): The North in the 1980's: new times in the 'Great North' or just more of the same? Area, 23, 1, pp. 47-56

Hudson, R. (1992a): The Japanese, the United Kingdom automobile industry and the automobile industry in the United Kingdom, Change in the Automobile Industry : an International Comparison. Discussion Paper No.9, University of Durham, Department of Geography

Hudson, R. (1992b): Institutional change,cultural transformation and economic regeneration : Myths and realities from Europe's old industrial areas. Occasional Paper No. 27, University of Durham, Department of Geography

Hudson, R. (1995): Is inward investment the key to the North's future economic development? In: Evans, L., Johnson, P. and Thomas, B. (eds.), The Northern Region: Progress and Prospects. Mansell, London (forthcoming)

Hudson, R. and Sadler, D. (1991): Manufacturing success? Reindustrialisation policies in Derwentside in the 1980's. Occasional Paper No. 25, University of Durham, Department of Geography

Hudson, R. and Williams, A. (1989): Divided Britain. Belhaven, London

Industrial Board of Sweden (1990): Leverantörer till fordonsindustrin. SIND 1990:2. Stockholm

Jacquemin A., Rainelli, M. (1984): Filières de la nation et filières de l'entreprise. Revue Economique 35, pp. 379-392

Jacot, J.H. (1980): Du fordisme au toyotisme? Les voies de la modernisation du systäme automobile en France et au Japon. La Documentation Francaise, Etudes et Recherches no. 7-8, Paris

Jalabert G. and Gregoris, M. (1987): Turin: de la ville-usine a' la technopole. In: Annales de geographie, no 538, pp. 680-704

Jonsson, O. (1989): Från Fordism till flexibel specialisering - Mot ett nytt produktionskoncept. Mot ett nytt rum? Rapporter och notiser 93. Institutionen för kulturgeografi och ekonomisk geografi vid Lunds universitet, Lund

Jürgens, U. (1991): The changing contours of work in the car industry. Wissenschaftszentrum Berlin discussion paper FS II 91-202, Berlin

Jürgens, U., Malsch, T. and Dohse, K. (1989): Moderne Zeiten in der Automobilfabrik. Strategien der Produktionsmodernisierung im Länder- und Konzernvergleich Springer, Berlin

Kapitány, S. and Kállay, L. (1991): The motor industry of Eastern Europe : prospects to 2000 and beyond. Economist Intelligence Unit Special Report 2142, London

Kinch, N. (1992): Entering a tightly structured network - Strategic Visions or Network Realities. In: Forsgren, M. & Johanson, J. (eds.), Managing Networks in International Business. Philadelphia: Gordon and Breach

Krafcik, J. F. (1988): Triumph of the Lean Production System. In: Sloan Management Review, Autumn

Kowalski, R. (1992): Zur Leistungsfähigkeit der enhemaligen DDR-Automobilindustrie und zu Problemen ihrer Umgestaltung. In: Deiß, M, and Döhl, V. (eds), Vernetzte Produktion. Automobilzulieferer zwischen Kontrolle und Autonomie Campus, Frankfurt/M, pp. 247-76

Krumme, G. (1981): Making it abroad : The evolution of Volkswagen's North American production plans. In: Hamilton, F.E.I. and Linge, G.J.R. (eds), International Industrial Systems (Spatial Analysis. Industry and the Industrial Environment. Progress in Research and Application II). J. Wiley & Sons, Chichester, pp. 329-56

Laux, J. (1992): The European automobile industry. Twayne publishers, New York

Leborgne D, Lipietz, A., (1989): New technologies, new modes of regulation: some spatial implications. Society and Space 6, pp. 263-80

Lehmann, J.P. (1992): France, Japan, Europe and industrial competition : the automotive case. International Affairs 68, pp. 37-53

Lipietz, A. (1992): The regulation approach and capitalist crisis: an alternative compromise for the 1990s. In: Dunford, M. & Kafkalas, G. (eds.), Cities and regions in the new Europe: the global-local interplay and spatial development strategies. London: Belhaven Press, pp. 309-34

Lompe, K., Müller, T., Rehfeld, C. and Blöcker, A. (1991): Regionale Bedeutung und Perspektiven der Automobilindustrie. Die Beispiele Südostniedersachsen und Südhessen (Graue Reihe Neue Folge 34). Hans Bückler Stiftung, Düsseldorf

Ludvigsen Associates (1988): Research on the 'Cost of non-Europe' basic findings volume 11 : the EC 92 automobile sector. Commission of the European Communities

Lundvall, B. A. (ed) (1992): National systems of innovations. Towards a theory of innovations and interacting learning. Pinter, London

Lung, Y. (1994): Modèles industriels et géographie de la production. In: Rallet A, Torre, A. (eds), Economie industrielle - Economie spatiale. Economica, Paris (forthcoming)

Lung, Y. and Mair, A. (1993): Innovation institutionelle, apprentissage organisationel et contrainte de proximité : Les enseignements de la géographie du juste à temps. Revue d'Economie Régionale et Urbaine 3, pp. 387-403

Lungwitz, R. and Kreißig, V. (1992): Sozialer und wirtschaftlicher Wandel. In: der Automobileindtrie der deuen Bundesländer. in Heidenreich, M. (ed), Krisen, Kader, Kombinate. Edition Sigma, Berlin, pp. 173-85

Mair, A. (1991): Just-in-time manufacturing and the spatial structure of the automobile industry, Change in the Automobile Industry : An International Comparison. Discussion Paper No. 4, University of Durham, Department of Geography

Mair, A. (1992): Just-in-time manufacturing and the spatial structure of the automobile industry: Lessons from Japan. In: Tijdschrift voor Economische en Sociale Geografie 83, 2, pp. 82-92

Marceau, J., Jureidini, R. (1992): Giants and dwarfs : Changing technologies and productive interlinkages in Australian manufacturing industry. In: Marceau J (ed), Organisations, technologies, and cultures in comparative perspective. de Gruyter, Berlin, pp. 151-170

Mehl, R. (1993): Fiat Auto : Struktur schlägt Strategie. Aufstieg und Fall eines autoritär gerührten Unternhmens. Edition Sigma, Berlin

Meinig, W. (1993): Partnerschaft im Interaktionsfeld Automobilhersteller - Zulieferer. In: Meinig, W. (ed), Wertschöpfungskette Automobilwirtschaft. Gabler, Wiesbaden, pp. 203-21

Mendius, H.G. and Wendeling-Scröder, U., (eds) (1991): Zulieferer im Netz. Neustrukturierung der Logistik am Beispiel der Automobilzulieferung. Bund-Verlag, Köln

Ministry of Industry (1990): Svensk industri och industripolitik 1990. Stockholm: Industridepartementet

Mitsubishi Research Institute (1990): The strategies of Japanese Automakers in Europe. mimeo, Tokyo

Momiglanno, F., (a cura di) (1986): Le leggi della politica industriale in Italia. Il Mulino, Bologna

Morris, J. (1992): Flexible specialisation or the Japanese Model: Reconceptualising a new regional order. In: Ernste, H, and Meier, V. (eds), Regional Development and Contemporary Industrial Response. Extending Flexible Specialisation. Belhaven, London, pp. 67-80

Moscini, A. and Rullani, E. (1978): Il gruppo nello sviluppo dell'impresa industriale. Con an'analisi del caso Fiat. Isedi, Milano

National Audit Office (1989): Department of Trade and Industry : Sales of Rover Group plc to British Aerospace plc. HMSO, London

National Economic Development Office (1991): The Experience of Nissan Suppliers: Lessons for the UK Engineering Industry. Millbank Tower, London SW11 4QX

Noble, D. (1992): Industry Dynamics in Europe: The Motor Industry. In: Hamill, J. et al. (eds), Europe and the Multinationals. Aldershot, Edward Elgar Publishing

Official Journal of the European Communities (1989): Community framework on state aids to the motor vehicle industry 89C 123/03

Okamura, C., Kawahito, H. (1990): Karoshi. Mado-Sha, Tokyo

Olberding, S. (1990): Die kunststoffverarbeitende Industrie im Landkreis Vechta. In: Klohn, W. (ed), Exkursionen zur Siedlungs- und Wirtschaftsgeographie in Nordwestdeutschland (Vechtaer Studien zur Angewandten Geographie und Regionalwissenschaft 1). Vechtaer Druckerei und Verlag, Vechtra, pp. 183-198

Organisation for Economic Cooperation and Development (1990): Industrial Policies in OECD Countries : Annual Review 1990. OECD, Paris

Oxford Motor Industry Research Project (1990): Cowley Works: Why it matters, why it must be saved. Oxford

Paché, G. (1991): L'Impact des Stratégies d'Entreprises sur l'Organisation Industrielle: PME et Reseaux de Compétences. Revue d'Economie Industrielle, No. 56, pp. 58-70

Pearce, M.C. (1987): Joint Ventures and Agreements in the Western Europe Motor Industry. The Economist Intelligence Unit. Automobile Special Report 7, London

Pemberton, M. (1991): Europe's motor industry after 1992 : a review of single market legislation and its implications. Economist Intelligence Unit Special Report 2090, London

Perroux, F. (1955): Note on the Concept of 'Growth Poles'. In: Livingstone, I (ed), Development Economics and Policy. London, 1981

Platzer, H. W. (1991): Gewerkschaftspolitik ohne Grenzen? Die transnationale Zusammenarbeit der Gewerkschaften im Europa der 90er Jahre. Dietz Nachf, Bonn

Porter, M.E. (1990): The Competitive Advantage of Nations. London and Basingstoke, Macmillan

Rawlinson, M., Wells, P. (1993): Japanese globalisation and the European automobile industry. Tijdschrift voor Economische en Sociale Geografie 84, pp. 349-361

Rehfeld, D. (1992): Beziehungen zwischen Branche, Konzern und Region in der Automobilindustrie. In Deiß, M., and Döhl, V. (eds), Vernetzte Produktion, Automobilzulieterer zwischen Kontrolle un Autonomie. Campus, Frankfurt/M, pp. 209-37

Reich, S. (1989): Roads to follow: regulating direct foreign investment. International Organisation 43, pp. 543-84

Rhys, G. (1979): The components sector. In Bhaskar, K. (ed), The future of the UK motor industry. Kogan Page, London, pp. 303-24

Rossignolo, G.M. (1971): L'industria automobilistica nell'economia italiana: un'analisi input-output. L'impresa, March-April

Sabel, Ch, Kern, H., Herrigel, G. (1991): Kooperative Produktion. Neue Formen der Zusammenarbeit zwischen Endfertigern und Zulieferern in der Automobilindustrie und die Neuordnung der Firma. In: Mendius H.G. Wendeling-Schröder, U. (eds), Zulieferer in Netz. Neustrukturierung der Logistik am Beispiel der Automobilzulieferung. Bund-Verlag, Köln, pp. 203-227

Sadler, D. (1991a): Strategic change in the west European automotive components industry. Change in the Automobile Industry : An International Comparison. Discussion Paper No.8, Department of Geography, University of Durham

Sadler, D. (1991b): Beyond 1992 : the evolution of European community policies towards the automobile industry. Change in the Automobile Industry : An International Comparison. Discussion Paper No.7, Department of Geography, University of Durham

Sadler, D. (1992a): The global region : production, state policies and uneven development. Pergamon, Oxford

Sadler, D. (1992b): European Community industrial policy : strategic deficits and regional dilemmas. Environment and Planning A

Sadler, D. (1992c): A comparative analysis of corporate performance in the automotive assembly and components industries, 1987-91. Change in the automobile industry. Discussion Paper No. 14, Department of Geography, University of Durham

Sadler, D. (1994): The geographies of "Just-in-Time" : Japanese investment and the automotive components industry in Western Europe. Economic Geography: 70, pp. 41-50

Sallez, A. and Schlegel, J. (1963): La sous-traitance dans l'industrie. Dunos, Paris

Saraceno, P. (1986): I trasferimenti sociali, in Artoni R., Pontarollo E. (eds), Trasferimenti, domanda pubblica e sistema industriale. Il Mulino, Bologna

Savary, J. (1990): Les Implantations Industrielles Étrangeres dans le Sud-Ouest de la France: Bilan et Perspectives. Revue Géographique des Pyrénèes et du Sud-Ouest, Tome 61, Vol. 1, pp. 85-111

Savary, J. (1991a): European Integration and the Globalization of Ford Strategy in Europe: an Historical Analysis, (mimeo)

Savary, J. (1991b): Transnational Corporations, Globalization Process and the Future of Peripheral Regions (mimeo)

Schamp, E. W. (1991): Towards a spatial reorganisation of the German car industry? The implications of new production concepts. In: Benko G. and Dunford M. (eds), Industrial change and regional development: the transformation of new industrial spaces. Belhaven, London, pp. 159-170

Schamp, E. W. (1994): Financial services, leasing and regional development in the Federal Republic of Germany. In: Schamp, E.W., Linge G.I.R., and Rogerson Ch. (eds), Finance Institutions and Industrial Change - Spatial Perspectives. de Gruyter, Berlin (forthcoming), pp. 118-33

Schregle, J. (1989): Worker's participation in the Federal Republic of Germany in an international perspective. In: Gladstone A. et al. (eds), Current issues in labour relations: An international perspective. de Gruyter, Berlin, pp. 105-114

Scott, W. G. (1991): Innovazione e mercato nell'industria automobilistica. Isedi, Milano

Semlinger, K. (1990): Personalanpassung und Personalentwicklung in der deutschen Stahl und Automobilindustrie. Zum Einfluß industrieller Beziehungen und öffentlicher Regulierung Arbeitspapiere aus dem Arbeitskreis Sozialwissenschaftliche Arbeitsmarktforschung (SAMF), Paderborn

Semlinger, K. (1991): New developments in subcontracting : Mixed market and hierarchy. In: Amin, A., and Dietrich, M. (eds), Towards a New Europe? Structural Change in the European Economy. Edward Elgar, Aldershot, pp. 96-115

Silva, F., Grillo M. and Prati, M. (1982): Il mercato italiano dell'auto nel contesto europeo. Angeli, Milano

Sinclair, R. and Walker, D.F. (1982): Industrial development via the multinational corporation : General Motors in Vienna. Regional Studies 16, pp. 433-42

Sleigh, P. (1988): The UK automotive components industry. Economist Intelligence Unit Automotive Special Report 10, London

Sleigh, P. (1991): The European automotive components industry : 1991 edition. Economist Intelligence Unit Special Report 2107, London

Sölvell, Ö., Zander, I. and Porter, M.E. (1991): Advantage Sweden. Stockholm: Norstedts

Spriano, G. (1985): Technological Choices and Spatial Organisation of Production in the Fiat System. Geography Discussion Paper, no 12, London School of Economics, London

Storper, M. and Harrison, B. (1992): Flexibilité, Hiérarchie et Développement Régional: Les Changements de Structure des Systèmes Productifs Industriels et leurs Nouveaux Modes de Gouvernance dans les Annés 1990. In: Benko, G.; Lipietz, A. (Eds), Les Régions Qui Gagnent. Paris, PUF, pp. 265-291

Streeck, W. (1984): Industrial relations in West Germany: A case study of the car industry. Heinemann, London

Svimez (1992): Mezzogiorni d'Europa. Il Mulino, Bologna

Swain, A. (1992): Eastern Europe and the Global Strategies of Automobile Producers. Change in the Automobile Industry : An International Comparison. Discussion Paper No. 11 University of Durham, Department of Geography

Taylor, M, Thrift, N. (1982): Industrial linkage and the segmented economy. Environment and Planning A 14 : 1601-1632

The Guardian, various issues

The Economist (1992): Robert Bosch: braking, 25 April, pp. 97-8

Turner, L. (1991): Democracy at work. Changing world markets and the future of labour unions. Cornell University Press, Ithaca and London

Tyne and Wear Development Corporation (1991): Automotives - the Driving Force of the New North East. Scotswood Business Park, Newcastle

van Tulder, R., and Ruigrok, W. (1993): Regionalism, localisation or glocalisation : the case of the world car industry. In: Humbert M (ed), The Impact of Globalisation on Europe's Firms and Industries. Pinter, London, pp. 22-33

Veltz, P. (1992): Hiérarchies et Reseaux dans l'Organisation de la Production et du Territoire. In: Benko, G.; Lipietz, A. (Eds), Les Régions Qui Gagnent, Paris. PUF, pp. 293-313

VDA (Verband der Automobilindustrie e.V.) (1992): Tatsachen und Zahlen aud der Kraftverkehrswirtschaft, 56. VDA, Frankfurt/M

Volpato, G. (1991): Le systeme automobile italien. In: De Banville E., Chanaron J.J. (eds), Vers un systeme automobile europeen. CPE Economica, Paris

Walker, R. (1988): The geographical organization of production-systems. Environment and Planning D : Society and Space 6, pp. 377-408

Weißbach, H.J. and Weißbach, R. (1987): Logistiksysteme in der Automobilindustrie. Wissenschaftszentrum Berlin discussion paper IIVG/dp, pp. 87-215 Berlin

Wells, P. and Rawlinson, M. (1992): New procurement regimes and the spatial distribution of suppliers : the case of Ford in Europe. Area 24, 4, pp. 380-90

West Midlands Industrial Development Association (1989): Vehicle components sector report. Coleshill

Wickens, P. (1988): The Road to Nissan. Macmillan, London

Wilkinson, J., Morris, J. Oliver, N. (1992) Japanizing the world : The case of Toyota. In: Marceau J (ed), Organisations, technologies, and cultures in comparative perspective. de Gruyter, Berlin, pp. 133-150

Wilks, S. (1988): Industrial Policy and the Motor Trade. Manchester University Press, Manchester

Womack, J.P., Jones, D.T., Ross, D. (1990): The Machine that Changed the World. Maxwell Macmillan International, New York

Wortmann, M. (1990): Produktionsverlagerungen aus der Bundesrepublik Deutschland an die europäische Peripherie. Vierteljahresberichte, 122, pp. 381-92

Author Index

Subject Index

Location Index

Springer-Verlag and the Environment

We at Springer-Verlag firmly believe that an international science publisher has a special obligation to the environment, and our corporate policies consistently reflect this conviction.

We also expect our business partners – paper mills, printers, packaging manufacturers, etc. – to commit themselves to using environmentally friendly materials and production processes.

The paper in this book is made from low- or no-chlorine pulp and is acid free, in conformance with international standards for paper permanency.

FSC
www.fsc.org
MIX
Papier aus verantwortungsvollen Quellen
Paper from responsible sources
FSC® C105338

If you have any concerns about our products,
you can contact us on
ProductSafety@springernature.com

In case Publisher is established outside the EU,
the EU authorized representative is:
Springer Nature Customer Service Center GmbH
Europaplatz 3, 69115 Heidelberg, Germany

Printed by Libri Plureos GmbH
in Hamburg, Germany